Fallen Heroes

A Tribute From Fire Engineering

Photo by Time, Inc.

September 11, 2001

FALLEN HEROES

A Tribute From Fire Engineering

SEPTEMBER 11, 2001

HV
6432
.F354
2001

Copyright© 2001 by
PennWell Corporation
1421 South Sheridan
Tulsa, Oklahoma 74112
1-800-752-9764
sales@pennwell.com
www.pennwell-store.com
www.pennwell.com

Publisher: Margaret Shake
Editor: Jared Wicklund
Cover, Book Design and Prayer Illustration: Clark Bell
Marketing Coordinator: Julie Baxter Simmons
National Account Coordinator: Francie Halcomb
MultiMedia Coordinator: William Rufus Clarke
IPG Staff: Travis Wallace, Tim Adams, Gloria Laughlin

Library of Congress Cataloging-in-Publication Data Pending

All rights reserved. No part of this book may be reproduced, stored in a retrieval system, or transcribed in any form or by any means, electronic or mechanical, including photocopying and recording, without the prior written permission of the publisher.

Printed in the United States of America

1 2 3 4 5 05 04 03 02 01

SEPTEMBER 11, 2001

Photo by Time, Inc.

TABLE OF CONTENTS

Fire Engineering Courage and Valor Foundation vii

Acknowledgements ix

Dedication xi

Foreword 1

History of the Fire Department of New York 7

Our Most Tragic Day 17

On Scene at the Pentagon 29

Tributes and Reflections 35

    Oklahoma's Friend, Ray Downey 35

    Our Brother, Andrew Fredericks 39

    All That My Children Have Taught Me 41

    The Heart and Soul of Squad 18 43

    The World Reacts 47

In Memory of Our Fallen Heroes 51

Afterword 141

The Fireman's Last Call 151

Firefighter's Prayer 152

FALLEN HEROES

FIRE ENGINEERING COURAGE AND VALOR FOUNDATION

Photo by Time, Inc.

To: Americans Everywhere
From: Robert F. Biolchini, President and CEO,
PennWell Corporation
Subject: Fire Engineering Courage and Valor Foundation

The events of September 11, 2001 changed our lives and our nation. Each of us has been affected in some personal and private way. Yet we all share the grief of the families and friends of our Fallen Heroes–the firefighters and rescue personnel who lost their lives that fateful day.

Knowing that the shock and sadness we have felt following this tragedy, and all of its subsequent events, may eventually fade, PennWell decided it was important to establish a memorial that would exist for many years into the future.

Accordingly, the corporation has set up The Fire Engineering Courage and Valor Foundation–a tax-free foundation that will present a monetary award to the firefighter (or his/her family) who has exhibited exemplary courage and valor in a rescue operation during the preceding year, in memory of those who sacrificed their lives on September 11. The Fire Engineering Courage and Valor Award will be presented annually at the Fire Department Instructors Conference & Exhibition in Indianapolis, with the first presentation during FDIC 2002.

Fire Engineering magazine dates back 124 years to its start in 1877. Both Fire Engineering and FDIC specialize in training the firefighters who serve us in disasters large and small. The staff personally knew and worked with many of these heroes–nine of them were contributing authors and instructors.

All profits from "Fallen Heroes" will be contributed to the Foundation, with matching contributions from the Corporation. Our goal is to raise and fund the Foundation with $1,000,000.

Thank you for helping preserve the memory of September 11, 2001 and for aiding the cause to honor our firefighter heroes for many years to come.

(Additional contributions may be sent to The Fire Engineering Courage and Valor Foundation, c/o Valley National Bank, 8080 South Yale, Tulsa, OK 74136.)

SEPTEMBER 11, 2001

Photo by Time, Inc.

ACKNOWLEDGEMENTS

The publisher and staff would like to express sincere sympathy to all those touched by the September 11 tragedy and acknowledge the following for their contributions to this book:

Don Bowden, Associated Press
Garry Briese, IAFC Executive Director
John Ceriello, FDNY
Firefighter Jay Comella, Oakland (CA) Fire Dept.
Battalion Chief Ted Corporandy,
    San Francisco (CA) Fire Dept.
Chief Brian Dixon, FDNY
FEMA News Photos
Tony Greco, Photographer
Ron Jeffers, Photographer
Janet Kimmerly, WNYF Editor, FDNY
Robert P. Mitts & Richard Smulczeski,
    www.FdnyPhotography.com
John P. Melfa, Photographer

Jennifer Narcisco, Student
1st Asst. Chief Vincent Narcisco,
    Valley Cottage (NY) Fire Dept.
Battalion Chief John Norman, FDNY
Oklahoma Governor Frank Keating
Cynthia Pelzner, Time, Inc.
Mark Seliger & Shelter Serra, Seliger Studio, New York
Steve Spak, Photographer
U.S. Department of Defense (www.defenselink.mil)
Captain Michael Veseling,
    City of Naperville (IL) Fire Dept.
Firefighter Geoff Williams,
    Central Scotland Fire Brigade

FALLEN HEROES

"O horror! horror! horror! Tongue nor heart cannot conceive nor name thee!"
William Shakespeare

Photos by Time, Inc.

DEDICATION

This book is dedicated to the Fallen Heroes of September 11, 2001. A hero is defined as a person noted for feats of courage or nobility of purpose, especially one who has risked or sacrificed his or her life. Attributes include courage, valor, gallantry. Who could deny that the fire and emergency personnel of that horrible, fateful day exhibited all of the above, and more?

Theirs was not an ignorant nor inconsiderate dash to danger – but a rush to rescue innocents from disaster, carried out with the confidence that comes from training and experience. We owe them; America owes them. They are truly heroes, fallen perhaps, but never to be forgotten.

Margaret Shake, Publisher

SEPTEMBER 11, 2001

FOREWORD

By Bill Manning, *Fire Engineering* Magazine

THE WORLD WAS FOREVER CHANGED ON TUESDAY, SEPTEMBER 11, 2001. THE FIRE SERVICE WILL NEVER BE THE SAME.

I WAS AT MY DESK MAKING TRAVEL PLANS. PLANS FOR THE FUTURE. LOOKING AHEAD. SOMEONE RAN INTO THE OFFICE. A PLANE HAD JUST CRASHED INTO ONE OF THE TWIN TOWERS. WE GRABBED A TV AND SET IT UP IN MY OFFICE.

FALLEN HEROES

By this time our whole staff was in the room, horrified, wondering, among many things, how that plane got there and how the fire department would conduct the operation.

Suddenly, a large plane came into view on the screen, headed for the towers. It banked to the left. The television screen went blank. We were under attack.

Two of us hopped in a car and headed south, listening to radio reports along the way. The Pentagon was hit. What was next?

We drove about 20 minutes to a hill in New Jersey, a few miles from the city, that offered a breathtaking view of the Manhattan skyline. From there, with a small group of onlookers, we watched in utter disbelief as the Twin Towers fell. It seemed impossible but it was real.

Like many, we wept.

Throughout the day, we gathered information from the media, tried to call loved ones, made a lot of consolation calls. Spoke a lot of angry words. We knew the time of the incident. We knew some 50,000 people worked in the towers alone. We knew our brothers in the New York City Fire Department would be in there, meeting the situation head on, with selflessness and determination, in the end, as in the beginning, in a virtual circle of lifegiving and hope, helping others.

Photo by John P. Melfa

Photo by Time, Inc.

SEPTEMBER 11, 2001

Photo courtesy of FEMA

Photo courtesy of FEMA

Photo courtesy of Fire Engineering

The following evening we made it across the bridge into a virtual ghost town. The smell of the collapse was everywhere.

The lower two miles of the island were completely shut down except for those involved in the response and a few hardy residents who chose to tough it out. Making it through initial barricades, walking southward into a hell that used to be the high-rise canyons of the financial center. Each step brought us closer to a scene that resembled a war zone.

Fire vehicles, police vehicles, National Guard trucks, heavy equipment, service trucks of every imaginable type, rehab and food stations, all stretched out along West Street as far as the eye could see. Police directed trucks through the eye of a needle to get resources to the site. Amidst lines of National Guardsmen and federal workers and medical personnel in their

FALLEN HEROES

Photo by Richard Smulczeski, FdnyPhotography.com

Photo courtesy of FEMA

scrubs, amidst a wash of utility employees and tired firefighters covered in white dust, I noticed a resident walking her dog.

Great clouds of dust lifted up into the evening sky from behind the many buildings that hid the view of the collapse site. Word was passed that nonessential emergency workers were to stand by outside an eight-block-radius due to secondary collapse threats at two other large buildings adjacent to what used to be the towers.

We hitched a ride with a fire equipment manufacturer providing services to one of the FEMA USAR task forces. We were now two blocks from Ground Zero.

Those two blocks were long in my mind. I was overwhelmed with a sense of dread and confusion. As I walked, to the left of me, as my boots crunched bits of debris, firefighters and other workers moved along a chow line, silhouetted in the darkness by the eerie light cast from the collapse site. A little further along, a machine with giant steel forceps grabbed crushed and burned cars and removed them from the street.

"We had a clear view of the murder site, with the smoke rising over the New York horizon, 15 miles away. It was beyond words." Billy Goldfelder, Battalion Chief, Loveland-Symmes (OH) Fire Department

September 11, 2001

We were now just one block east of West and Vesey streets. Sixty- to 100-foot tall piles of what was once the World Trade Center hissed and steamed. Aerial master streams wet them down. Firefighters had set up a small rest area with tables and chairs from the shell of the building. Someone erected a small, handwritten sign: "Steve's Café. Today's menu: 1. Water 2. Water 3. More Water."

Ground Zero slams the mind with the onrush of the gruesome and surreal: Directly in front of you, 100 stories of the North Tower collapsed, pancake fashion, into an incomprehensible pile of steel and concrete. To the east, the south, the west: Twisted, hissing remains that once held so much life. From the northern perspective at night, it was impossible to grasp how far the piles extended southward, but it seemed like they stretched all the way to hell. Giant steel structural members stuck out of the pile in every direction. A billion broken pieces, a billion shattered dreams.

The air was charged with the dust of death and the sound of heavy equipment hard at work. Huge cranes picked off pieces of building from the top of the collapsed pedestrian walkway

> "It's as if a volcano erupted, a tornado struck, and a bomb fell all at the same time. The scene doesn't seem real; it's like a movie—surreal, a word used by so many to describe this horrible atmosphere." **Brian Hendrickson, Oakland (NJ) Fire Department**

Photo by Time, Inc.

FALLEN HEROES

"Now I can relate to people in foreign countries who have had their homeland bombed, what Germany looked like when it was bombed. Now I can relate to how the devastation looked." Jim King, Firefighter, North Arlington (NJ) Fire Department

Photo courtesy of FEMA

Photo by Time, Inc.

on West Street. Adjacent to them, firefighters combed a 30-foot-high pile of debris, passing five-gallon-buckets by hand. The command post was abuzz with activity. I could not help but think of my friend Ray (Downey, Chief of Rescue Operations, FDNY) and more than 300 of his hero brothers.

Teams of weary but determined firefighters streamed through the muck to and from the avenue to hell. One firefighter, tired of face, was saddened by the fact that they had recovered only one live person that day.

Gritty. Determined. Ready. These are thoughts that passed through my mind as I watched this surreal play, the aftereffects of the unthinkable. But this is, after all, the nature of firefighters and our fire service. And at that moment it was so abundantly clear, in the eyes of the relentless firefighters at the scene, that the brothers and sisters who perished in the World Trade Center live on in the hearts and minds of firefighters across this great country, immortalized in the acts of courage, large and small, every day. They live on, friends, carried on the wings of your greatness. And that brings a sweetness to our bitter tears. They are alive in you.

Photo by Time, Inc.

HISTORY OF THE FIRE DEPARTMENT OF NEW YORK

ORGANIZED FIREFIGHTING BEGAN IN NEW YORK IN 1648 WHEN PETER STUYVESANT, GOVERNOR OF NEW AMSTERDAM, APPOINTED FOUR FIRE WARDENS.

IN A STROKE OF POLITICAL SAVVY, HE NAMED TWO DUTCHMEN AND TWO ENGLISHMEN TO FILL THE NEW POSTS. THESE ORIGINAL FIRE WARDENS WERE ALL ABLE AND HONEST CITIZENS AND POLITICALLY CORRECT PUBLIC SERVANTS OF THEIR TIME. FINES LEVIED FOR DIRTY CHIMNEYS PROVIDED FUNDS FOR THE MAINTENANCE OF BUCKETS, HOOKS, AND LADDERS.

A fire watch was also established, requiring each male citizen to take his turn in participating.

The fire wardens were empowered to visit every house, inspect all chimneys and see that they were swept clean, and to ensure that no wooden chimneys were built. A fine of up to 25 guilders could be imposed if a house caught fire due to negligence or if the flames extended from the fireplace.

In 1657, New Amsterdam incorporated and streets were given names. Bucket brigades were formed and equipped with leather buckets made by Dutch shoemakers of the colony. The new fire wardens designated a group of eight men to walk the newly paved streets after dark, watching for fires. They carried large wooden rattles to sound an alarm if a fire was spotted.

Seven years later New Amsterdam became New York, and the rattles gave way to bells. By 1697, four men known as the "Night Watch" were walking the streets with bells at the ready.

September 11, 2001

(Above) A fire officer leads a steamer and a ladder truck through the crowded streets.

(Right) The shed that housed Engine 3 on Nassau St., across from City Hall, in 1801.

They rang their bells and announced the time and weather every hour.

It was not until 74 years later, in 1731, that fire brigades were put into service. Two hand-drawn pumpers, brought from London on the good ship Beaver, were the first fire engines in the the colony. They were designated as Engine Company 1 and Engine Company 2, manned by appointed volunteers who were to be "strong, able, discreet, honest and sober men." They were responsible for the "care, management, working and use of the fire engines and other tools and instruments of fires" and "ready at a call by night and by day." If they failed to answer an alarm "without reasonable cause," they would be fined 12 shillings. For their effort, they would be exempt from jury duty, serving in the militia, and other mundane duties faced by the average citizen.

The city's first firehouse was built in 1736, in front of City Hall on Broad Street, between Exchange Place and Wall Street. The apparatus protected a city of 1,200 houses, with a population approaching 9,000. Late in 1743, the first successful American-made apparatus was added to the department. Thomas Lote, a cooper and boat builder, covered much of the box of his hand pumper in brass and relied on hand-operated levers only. Engine Number 3 soon went into service and became known unofficially as "Old Brass Backs."

The first hook and ladder company was placed into service in 1774. Two years later, the department had grown to 170 men, assigned to eight engine companies and two truck companies. Then the war with England closed in on the city.

General Washington fortified New York with troops and awaited the arrival of the British, led by General Sir William Howe. After the August 27-29 Battle of Long Island, General Washington was forced to evacuate. Virtually the entire fire department enlisted with Washington and marched away with the army, leaving their beloved fire engines behind. Six days after the firemen left the city, shouts of fire filled the late-night air. Flames were seen burning through the roof of the Fighting Cocks Tavern, which soon spread to six adjacent structures. But no alarm was sounded as all the bells in the city had been

FALLEN HEROES

New York Firemen worked through the night and into the next day before The Great Fire of 1835 could be contained.

carried off to be melted into munitions. Neither the soldiers nor the townspeople had any real idea how to effectively battle the flames. By the time the fire was over, approximately one-quarter of the city – nearly 500 buildings – had been destroyed.

When the firemen returned in 1783, they found block after block of charred remnants of buildings burned in the great fire. At their firehouses they found only two engines working and the department in shambles. Within a month, all the engines were repaired and orders were placed for hundreds of new buckets. Two years later the department was reformed by returning veterans, and the custom of officially naming companies was started, with Number 1 taking the name Hudson Engine Company No. 1, in honor of the explorer, Henry Hudson.

The Fire Department of the City of New York was now official, but it was staffed by volunteers until after the close of the Civil War. Fires became more frequent, as did deaths of the volunteering firemen. There were bitter rivalries between companies that hampered the effectiveness of the firefighting efforts. An outbreak of cholera that swept the city in 1832 forced the Chief Engineer, James Gulick, to hire horses to pull the engines to the fires because so many volunteers were too ill to respond.

Leather Stovepipe hat designed by Jacobus Turck around 1731 and improved leather helmet by Henry Gratacap in 1828.

By 1835 New York City was populated by a quarter of a million people. The city boasted 44 schools, more than 400 street lights, five theaters, about 35,000 houses – but only 1,500 firemen.

The introduction of the steam engine spelled doom for the volunteer department. The steam apparatus eliminated the need for men to pump the water. The horses had already ended the problem of hauling fire engines by hand. But the change to a paid department created resentment and resulted in rough and tumble battles fought on both personal and political levels.

At the beginning, the paid fire service covered only Manhattan (sections of Manhattan North were still protected by volunteers). The Act of 1865 united Brooklyn and New York cities to form a Metropolitan District, under an organization known as the Board of Metropolitan Fire Commissioners. There were four commissioners, appointed by and reporting to the Governor of New York State.

The first of the professional units, Engine Company No. 1, went into service on July 31, 1865. Members of the volunteers

The original members of Rescue Company 1, circa 1915.

were given preference over others in filling the rolls of the new paid department. By the end of 1865 the paid force consisted of an Assistant Engineer, 13 Engineers (Battalion Chiefs), 34 Engine Companies, and 12 Ladder Companies.

At this time, bellringers in bell towers spread throughout the city watched for fires, manning each tower in three-hour shifts. When a fire was discovered, they would telegraph the location of the fire to the Central Office in City Hall, and sound the alarm by striking the district number on the tower bell.

The Central Office then would telegraph the other tower bellringers, who would pick up the alarm for about 10 minutes. As alarm-receiving systems were installed in firehouses, the need for and use of these towers were eliminated.

The first rules of the Metropolitan Fire Department prescribed the number of men who might ride on the apparatus and specifically prohibited the Company Commanders from riding either to or from fires. Officers in command were to precede their apparatuses in going to or returning from a fire – and only the drivers were allowed to ride back to the Engine Houses.

Most of the early rescues involved ladders. With no breathing equipment available, and the normal delay in waiting for the steamer to get up enough pressure to charge the hose lines, the most effective efforts were over ladders. The standard ladder truck was pulled by two horses and carried wooden portable ladders, both extension and straight, in sizes up to 73 feet. The Company Commander was required to keep at least

(Above) The Triangle Shirtwaist fire of March 25, 1911.

(Left) The Standard Oil Company fire in Greenpoint, Brooklyn, September 13, 1919.

SEPTEMBER 11, 2001

(Right) A five alarm rag shop fire on East 5th Street, January 12, 1924.

(Below) A survivor of The Triangle Shirtwaist fire is rushed to a waiting ambulance, March 25, 1911.

nine men in quarters at all times, to ensure adequate manpower to raise the ladders.

Just three years after the start of the paid force, the Metropolitan Fire Department began a system of recording the heroic deeds of its members. The first name placed on the new Roll of Merit was Assistant Foreman Minthorne Tompkins of Ladder Company 1, who climbed an extended ladder into black smoke to rescue a woman trapped in a hotel by flames – then went back into the building to help safely remove another six people.

By 1898 the various areas of the city were consolidated under the unified command of the first Commissioner of the Fire Department. The last volunteer unit was disbanded in 1928.

One of the most versatile of firefighting tools was named for a New York City Fireman. Hugh A. Halligan served the FDNY for more than 25 years, appointed First Deputy Fire Commissioner in 1941. But his most important contribution was the refinement of the forcible entry tool – a claw tool that started out heavy and basically "off-centered," as dangerous to the man holding it as to the doors it was aimed at. Halligan studied every curve, corner and dimension, and ended up with a tool that could be held in one hand, that wouldn't chip or

Fallen Heroes

Photo by Mark Seliger

break, wouldn't tire out the user, and could be used safely, with full efficiency. A modern version of the Halligan is still used by the fire service today.

Members of the FDNY not only covered fires and city emergencies with bravery and daring, but through the years have served proudly in our country's military: 275 firemen joined up during the four-month Spanish-American War and 600 men from FDNY saw action in World War I. Fireman William L. Willis was awarded the Navy Cross for his extraordinary heroism in WWII. Fireman James Smith received the Congressional Medal of Honor while serving with the Navy in China. Firefighters from across the metropolitan area returned from abroad with many awards and Purple Hearts, all of them heroes.

Today the FDNY consists of more than 11,400 fire officers and firefighters, plus 2,800 emergency personnel, responsible for the protection of more than 8,000,000 residents. As they demonstrated on September 11, 2001, they are truly heroes.

"No man is worth his salt who is not ready at all times to risk his body, to risk his well-being, to risk his life, in a great cause." Martin Luther King, Jr.

SEPTEMBER 11, 2001

Photo by Time, Inc.

ORIGIN OF THE MALTESE CROSS

When a courageous band of crusaders known as the Knights of St. John fought the Saracens for possession of the Holy Land, they encountered a new weapon unknown to European warriors. It was a simple, but horrible device of war, imposing excruciating pain and agonizing death upon the brave fighters for the cross.

The Saracen's weapon was – Fire.

As the Crusaders advanced on the walls of the city, they were struck by glass bombs containing naphtha. After saturating the attackers with this highly flammable liquid, the Saracens hurled a flaming torch into their midst. Hundreds of knights were burned alive. Others risked their lives to save their brothers-in-arms from dying painful, fiery deaths.

Thus, these Crusaders became our first firemen, the first of a long list of courageous heroes. In recognition of their heroics, they were awarded a badge of honor, a cross similar to the one firefighters wear today. Since the Knights of St. John lived for centuries on the Mediterranean island of Malta, the cross came to be known as the Maltese Cross.

The Maltese Cross is a symbol of protection. It means that the firefighter wearing it is willing to lay down his life for you – just as the Crusaders sacrificed their lives for their fellow man so many years ago. The Maltese Cross is a fireman's badge of honor, signifying that he works in courage – a ladder rung away from death. The symbol of a hero.

15

FALLEN HEROES

"A man does what he must – in spite of personal consequences, in spite of obstacles and dangers and pressures – and that is the basis of all human morality."
John F. Kennedy

SEPTEMBER 11, 2001

OUR MOST TRAGIC DAY

By Bill Manning, *Fire Engineering* Magazine

IT WAS THE MOST TRAGIC, HORRIFIC DAY IN THE 250-YEAR HISTORY OF THE AMERICAN FIRE SERVICE. NEVER BEFORE HAVE SO MANY FIREFIGHTERS FALLEN IN ONE EVENT.

AT 08:45 ON SEPTEMBER 11, 2001, THE FIRST OF FOUR HIJACKED PLANES, AMERICAN AIRLINES FLIGHT 11, CARRYING 92 PASSENGERS AND CREW MEMBERS, SLAMMED INTO TOWER 1 (NORTH TOWER) OF THE WORLD TRADE CENTER, A SYMBOL OF AMERICAN FREEDOM AND CAPITALISM.

FALLEN HEROES

The Fire Department of New York immediately dispatched a five-alarm assignment, bringing more than 200 personnel to the scene.

At 09:03 the second hijacked plane, United Airlines Flight 175, carrying 65 passengers and crew, smashed into Tower 2 (South Tower). FDNY dispatched a second five-alarm assignment, bringing many more firefighters to the World Trade Center.

As all Americans know through the extensive media coverage of the incident, the effects of the crashes were unimaginably devastating—enough to bring a great city and a great nation at least temporarily to its knees.

The two Boeing 767 planes carried approximately 24,000 gallons of jet fuel each. The impact of the collisions resulted in huge explosions and fireballs. Occupants on floors near the impact are presumed to have been incinerated. Subsequent fires, burning at an estimated 1,700°F or more, were enough to

September 11, 2001

Photos by Ron Jeffers

"The building started coming down, and everybody ran. It was like being on the beach with this big wave coming at you. Only this wave was concrete and steel and glass." FDNY Chief Tom McCarthy

bring the already weakened steel of the exterior-wall-supported structures to the point of total and catastrophic collapse. Windows on upper floors were shattered by the impact, though some may have been broken by occupants. Either way, as the upper floors quickly became contaminated with smoke from the fires, many occupants fled to the windows in search of fresh air. Many people fell to their deaths—some perhaps pushed out in the panic, but others clearly choosing to jump to escape certain death on the fire floors. For those who could flee down stairwells, as well as for the firefighters assisting that effort, it was a race against time—though no one knew how much time they had.

It appears that FDNY units responding to the first plane collision in Tower 1 operated initially with the intention of attacking the fire—or operating handlines off the standpipe to

Photo by Time, Inc.

protect the stairwells and limit vertical extension—and evacuating occupants from at least some of the floors.

It is known that shortly after the first plane hit, occupants began self-evacuating not just Tower 1, but Tower 2 as well. Many had lived through and remembered the first attack on the World Trade Center in February, 1993. And it is believed that this rapid and widespread self-evacuation, while not the "recommended" occupant behavior in a "typical" high-rise fire, saved many lives on September 11. Surely, occupants realized from the incredible impact of the planes that this was an extraordinary event.

With the collision of Flight 175 into Tower 2, firefighters inside the buildings undoubtedly were operating in rapid evacuation mode. The elevators were not operational, so firefighters climbed the stairs to upper floors. It is impossible to know how many occupants within the WTC towers were saved that day by the firefighters—probably thousands. However, it is certain that firefighters operated with extreme courage and a steadfast sense of duty despite very difficult conditions—the hallmark of the fire service.

Likewise, without the benefit of communications traffic and company reports from that day, it is impossible to provide an analysis of strategic, tactical, and other critical factors within this enormous event. But it is certain that the incident command team understood the severity of the incident and operated with courage and leadership beyond expectation.

No one could have anticipated the collapse window for the Towers, given that, in the past, similarly constructed high-rise structures have withstood days' worth of heavy fire without total failure. Of course, this event was without precedent. But even if such a collapse window could have been predicted, occupants were still

September 11, 2001

"Fire trucks were squashed to the size of a desk. Firefighters were climbing piles, trying to listen, trying to find voids. Finding fatalities. When they carried out a firefighter, everyone stopped and took off their helmets." Raymond Kiernan, Fire Chief/Commissioner, City of New Rochelle (NY) Fire Department

evacuating at the time of collapse. Any second-guessing, in addition to being a grievous insult to the memory of firefighters who perished that day, would suggest that commanders had a choice. Having been rightly committed to interior operations, there was no choice but only one strategy: Get the people out and move to safe locations. And again, that irreversible strategy was carried out with unimpeachable courage, honor, and sheer heroism.

Time simply ran out for so many.

The deadly, catastrophic collapses of the towers—first Tower 2, under an hour after impact; then Tower 1, about $1^{1}/_{2}$ hours after impact—shook the country and the world. Undoubtedly, it was the worst disaster ever on mainland U.S. soil.

The count of dead or missing in the disaster has varied from 2,000 to 4,000. But exactly three hundred forty-three FDNY firefighters were caught in the collapse, including Chief of

"I've been in the fire service for 27 years. Being a marine in Vietnam is the closest I've come to this experience, but nothing of this magnitude, ever." Jim King, Firefighter, North Arlington (NJ) Fire Department

September 11, 2001

"In the last great attack on America, the attack on Pearl Harbor, the first casualties of that war were the members of our United States Navy," Mr. Giuliani said during the promotion ceremony at the MetroTech Center in Brooklyn. "They wore a uniform, like you do. In this war, the first large casualties are being experienced by the Fire Department of New York City."
Mayor Rudolph Giuliani

Department Peter Ganci, Deputy Commissioner William Feehan, and Fire Department Chaplain Father Mychal Judge. Entire companies from all city boroughs were decimated. Because the incident occurred during shift change, some companies responded with twice the usual number of personnel.

The immediate disaster area encompasses about seven square blocks of downtown Manhattan. Three large buildings completely collapsed—1, 2, and 7 World Trade Center. 3 World Trade Center (the Marriott Hotel) and 6 World Trade Center nearly completely collapsed. Five others have sustained significant structural damage. Prior to debris removal, the area contained 1.25 million tons of debris from the WTC alone. Engineers have estimated that the window glass from the Towers would cover 15 acres, there was enough crushed steel to build 20 Eiffel Towers, and the concrete would be enough to lay a five-foot-wide sidewalk for about a 450-mile stretch.

The towers collapsed in pancake fashion. If they had not, it is possible the death toll in this incident would have been even higher. Even so, the collapse fashion of the mammoth

FALLEN HEROES

Photo by The Associated Press

buildings—some 1,200 feet into about 120 feet (several stories above grade and several below)—created an extremely difficult and hazardous condition for search and rescue.

Almost immediately following the collapse, FDNY began a call-back procedure of off-duty personnel. A cloud of dust and thick smoke covered the disaster area and beyond. Fires were burning, including a nearby 10-story building that was fully involved. Debris was everywhere. Secondary collapse, both in and around the site, was a serious threat. Gas lines were thought to be in danger of rupturing. Fires burned—as they would for days—within the debris piles. Wounded littered the street. Lower Manhattan was a war zone.

After possibly being moved once or more in the hours following the collapse, the command post was positioned at the north corner of West and Vesey streets—just beyond the northwest corner of what had been the World Trade Center complex. The city's Office of Emergency Management, housed in the World Trade Center, evacuated successfully and set up its interagency command at Pier 92 along the Hudson River.

The governors of New York and New Jersey declared a state of emergency. And the federal government's presence was felt even prior to collapse, as fighter jets patrolled the air space.

"We have seen the state of our Union in the endurance of rescuers, working past exhaustion. We have seen the unfurling of flags, the lighting of candles, the giving of blood, the saying of prayers – in English, Hebrew, and Arabic. We have seen the decency of a loving and giving people who have made the grief of strangers their own."
President George W. Bush

September 11, 2001

Within hours, National Guard troops were sent to New York. City police and National Guardsmen attempted to secure the entire area of Lower Manhattan below 14th Street, with increasing security levels closer to Ground Zero.

President George W. Bush officially declared a National State of Emergency on September 14.

Initial search and rescue efforts were conducted by the fire department, police department, Port Authority police, emergency medical service agencies, and other agencies and individuals. City hospitals and hospitals across the Hudson River in New Jersey anticipated and mobilized for the treatment of large numbers of injured from the collapse site, but after the first few hours following the collapse, few live victims were found in the rubble who could be treated. Many thousands of people in Manhattan and beyond responded to calls for blood for banks that quickly became overstocked.

"I have been in the fire service 35 years, and I never saw anything even remotely like this and I hope I never see anything that tops it. I knew a few guys involved. I can only imagine what FDNY feels like losing hundreds they knew." Raymond Kiernan, Fire Chief/Commissioner, City of New Rochelle (NY) Fire Department

New York City firefighters—stunned by the possible loss of so many brothers in this senseless act of violence—set about ferociously to find the missing. During the first days of the incident, many worked day and night, uninterrupted. Thousands of firefighters from the Tri-State metropolitan area and all over the country converged on New York to offer their assistance, forming, along with thousands of other workers at the site, long bucket removal and debris-passing lines. Numerous times in these first days of the incident, work was stopped temporarily for a variety of reasons, most notably for the threat of secondary collapse, removal of large structural members, and fires in the debris that hampered void search. On at least two occasions, on-scene reports of imminent secondary collapse sent hundreds of people on the site running for their lives. Those were false alarms.

Photo courtesy of Fire Engineering

Fallen Heroes

Photo by The Associated Press

Photo by Time, Inc.

By Thursday afternoon, September 13, the shell-shocked incident command team, which had lost so many of its leaders and was carrying so many burdens, had to take additional steps to ensure site control, accountability, and scene safety by limiting access to only those firefighters and departments officially deployed. Over the next week or so, participation in the rescue effort at Ground Zero, from a fire service perspective, was scaled back primarily to FDNY, FEMA USAR teams, two state USAR teams, and a few assemblages of fire department teams from New Jersey and Long Island.

During the two-week period following the disaster, FEMA deployed and rotated more than a dozen 62-member USAR teams to New York. These included teams from Ohio, Pennsylvania, Massachusetts, Maryland, Indiana, Utah, Washington, Arizona, Missouri, California, Florida, and Texas. FEMA also sent its USAR Incident Support Team. A state team from New Jersey and a team from Puerto Rico were requested early on. These teams, with search dog capabilities in addition to extensive technical rescue skills, were detailed to FDNY sector commanders in each of four rescue operations sectors on the site for void search and other duties. As the search efforts became more well-defined, USAR teams were released.

Photo by Time, Inc.

SEPTEMBER 11, 2001

"There was no light, just generators to light the fireground. We saw vehicles twisted and thrown all over the place—police and fire vehicles and ambulances."
John Lewis, Firefighter, Passaic (NJ) Fire Department and FDIC West Conference Coordinator

FDNY, in response to its crisis of personnel, made numerous temporary and permanent changes to account for the great losses to many companies, including promotions, reassignments, temporary station closures, and shift modifications. Some on-duty shifts were split to cover both the site and normal response areas simultaneously. More than a dozen New Jersey fire companies were stationed in the borough of Staten Island to help staff fire stations there. Fire departments from across Long Island sent companies to help cover FDNY fire stations in Queens and Brooklyn.

On September 15, Fire Commissioner Thomas Von Essen, in a tearful ceremony, appointed Chief Daniel Nigro as the new FDNY chief of department and promoted 168 FDNY members, some of whom were missing in the collapse.

The trade unions, with their heavy equipment and special expertise, played a vital role in the rescue effort. By accounts from the scene, as initial debris strata and voids were searched, the operation began to rely even more on the riggers and crane operators. Equipment at the site included a high-reach (125-foot) excavator, probably the largest in the country. Thousands of tons of debris have been removed from the site, though some officials estimate that it will take several months to remove it all.

As the debris removal process moved into the sublevel areas, there were concerns about the integrity of the basement retaining wall around what was the entire complex. This wall kept out the surrounding 23 acres of landfill—and behind it, the Hudson River.

Late in the second week of the operation, units from the New York-New Jersey Port Authority, assisted by the Jersey City Fire Department, attempted to access the site from the PATH (train) tunnel from the New Jersey side. This tunnel was flooded, presumably from water main breaks or firefighting water, and 6,000gpm pumps were deployed. Rescuers had hoped to find survivable voids at the B6 sublevel, the train level, of the WTC. However, according to reports, a huge pile of condensed debris had collapsed down to that level, preventing access. That, plus the skyrocketing carbon monoxide levels in the tunnel, forced units to return to the other side of the river.

More than a week into the operation, New York City Mayor Rudolph Giuliani began expressing doubts about the chances of finding live victims. Very few bodies were recovered intact, and it appears from the horrific impact of collapse that many people may never be found. For some families, the only hope of positive victim identification is through DNA sampling.

The arduous task of recovery will continue for a long time.

Photo by The Associated Press

SEPTEMBER 11, 2001

ON SCENE AT THE PENTAGON

Courtesy of the U.S. Department of Defense
www.defenseLINK.mil

THE DAY BEGAN WELL IN WASHINGTON. THE WEATHER WAS BEAUTIFUL, WITH TEMPERATURES BEGINNING IN THE 50s AND RISING TOWARD 70. THE SUN WAS SHINING WITH A GENTLE BREEZE. AT THE PENTAGON, IT WAS A DAY LIKE ANY OTHER. THE DAY BEFORE, DEFENSE SECRETARY DONALD RUMSFELD HAD DECLARED WAR ON BUREAUCRACY THAT STIFLED INNOVATION. *THE EARLY BIRD* WAS FILLED WITH STORIES COVERING THAT EVENT.

Then CNN started broadcasting video of the World Trade Center North Tower burning. At first, it was unclear whether the inferno in New York was the result of terrorism or just an accident. The second plane ramming into the trade center's South Tower removed all doubt. People in the Pentagon exchanged shocked glances.

At 9:38 a.m., it was their turn.

Photos courtesy of FEMA

American Airlines Flight 77 had taken off without incident from Dulles International Airport 20 miles away. It was well on its way to the West Coast when hijackers took control and circled back toward Washington. The plane flew low, following Columbia Pike in Arlington, VA, a major avenue that beelines from the suburbs to the Pentagon. Witnesses remember hearing the plane throttle up just before it hit the Pentagon at Wedge One and ignited. Smoke billowed over the building while people inside groped to escape. Where the jetliner hit looked like someone had taken a knife and sliced the building down to the ground. Pieces of the aircraft littered the heliport area, and the fire truck and vehicles normally posted there were burning in flames.

The next several hours were the worst of times, but they brought out the best in the people of the Pentagon. Inside the offices, co-workers helped each other out of their offices and out of the building. Leaders ensured their people were accounted for as the evacuation progressed. Defense Protective Service officers charged to the scene and started directing the evacuation. Other officers went into the building to carry people to safety. Small acts of sacrifice and courage lowered the death toll. Colonels, captains, sergeants, yeomen, and Department of Defense (DoD) civilians searched offices and called out to anyone who

September 11, 2001

"Today our nation saw evil, the very worst of human nature, and we responded with the best of America, with the daring of our rescue workers."

President George Bush

Fallen Heroes

Photo by Time, Inc.

might still be inside. They fought through blinding, choking black smoke and heat to reach their co-workers.

Outside, the Arlington County police and fire departments arrived and started securing the area and pumping water on the blaze. Doctors, nurses, and emergency medical technicians who happened to be driving by stopped and started setting up triage areas for those wounded in the attack. More equipment arrived from the City of Alexandria (VA), Fairfax County (VA), Washington, and the Maryland counties. Helicopters flew the most seriously wounded to nearby hospitals from a makeshift heliport set up close to the Navy Annex, a building complex near the Pentagon. People who had gathered around the site to help found themselves transporting those with minor injuries to area hospitals. These were people with cuts, bruises, sprains, and some smoke inhalation. Helicopters and ambulances transported the most seriously injured. More resources continued to pour into the area as time wore on. Fort Belvoir engineers specializing in urban search and rescue arrived at the Pentagon. So did FBI agents and Federal Emergency Management Agency officials. Arlington County set up a joint operations center at Fort Myer, VA.

Photo courtesy of the Department of Defense

SEPTEMBER 11, 2001

"In any moment of decision, the best thing you can do is the right thing. The worst thing you can do is nothing."

Theodore Roosevelt

Photo courtesy of the Department of Defense

Photo by The Associated Press

In other parts of the building unaffected by the blast, the evacuation went calmly, with people stopping to secure classified items and ensuring all members were out of an office before securing it. One area never closed: The National Military Command Center. Secretary of Defense Donald H. Rumsfeld and the senior leadership remained inside the facility and monitored DoD activities worldwide.

As a result of the attack, 125 people were killed or remain unaccounted for, not including the 64 passengers on the hijacked plane, according to DoD documents. At the time of this publication, 118 remains have been recovered and transported to Dover Air Force Base, Delaware, for identification.

* * *

The Pentagon was built in 1941 out of reinforced concrete. The Pentagon's Wedge 1 section, which along with Wedge 2 took the brunt of the hijacked airliner's impact, had just undergone improvements as part of a total building renovation slated for completion by 2012. It is laid out in five concentric pentagonal "rings," the "E" being the outermost and "A" the innermost. The jet cut the building like a knife. It did not penetrate all the way into the center courtyard, but did reach the "B" ring. The airliner crashed low and diagonally into the Pentagon's outside "E" ring limestone wall. Floor-to-floor and interconnected vertical steel beams, sturdier windows, and Kevlar armor panels used in the revamped exterior wall helped slow down the plane and mitigate effects of the explosion as the plane crashed through the Pentagon. The portion of the Pentagon to the right of the initial impact area collapsed, hampering firefighter search and rescue efforts until the situation

Photo courtesy of FEMA

Photo by Time, Inc.

stabilized. Fortunately, the area had held together long enough for many workers to escape.

The construction of the Pentagon, while a benefit in the building's stability, had contributed to the stubborn fire. The Pentagon's roof consists of a layer of masonry, followed by a layer of horsehair insulation, topped by heavy timber, topped by slate. The fire had been hard to control because it had ignited the wood and was traveling between the concrete and slate layers, spreading into unrenovated sections. Trench cuts were used in several places to stop fire spread. Pools of aviation fuel would ignite occasionally and create thick black smoke over the building. The fire was not "contained" until about eight hours after impact. The engine and ladder companies did excellent interior operations to cut-off extension of fire into other areas of building; fire sprinklers were not a factor in control of the fire due to the amount of fire, flammable liquid, and broken pipes. Some handlines needed to be extended up to 600ft to effectively reach the fire (involving plenty of hand jacking of both supply and attack lines). The firefighters did an outstanding job.

The Pentagon's chief renovation official stated that repairs on the building stemming from the September 11 terrorist attack might take more than three years to complete.

SEPTEMBER 11, 2001

From left: Chief Gary Marrs of the Oklahoma City Fire Department, Oklahoma Governor Keating, and Chief Ray Downey. The three men became acquainted after the Oklahoma City bombing in 1995.

Photo by Fire Engineering

OKLAHOMA'S FRIEND, RAY DOWNEY

By Governor Frank Keating

I FIRST MET RAY DOWNEY LATE ON THE NIGHT OF APRIL 19, 1995, IN FRONT OF THE BLEAK, BOMBED-OUT SKELETON OF THE MURRAH FEDERAL BUILDING IN OKLAHOMA CITY. IT WAS RAINY AND COLD, AND WE WERE JUST BEGINNING THE AWFUL WORK OF BRINGING OUT THE DEAD FROM THE SCENE OF WHAT WAS, UNTIL SEPTEMBER 11, THE WORST DOMESTIC TERROR ATTACK IN AMERICAN HISTORY.

FALLEN HEROES

Photos courtesy of FEMA

For the next two weeks, as a member of New York's Urban Search and Rescue team and as site operations chief for the Federal Emergency Management Agency, Chief Downey was far more than a helping hand. He was an inspiration of unrivaled expertise. He was also a man whose beaming smile lightened some very dark hours. I am honored to say he became my friend.

On September 11, Chief Ray Downey rode the collapsing World Trade Center down to his death. We in Oklahoma fear that many more of our friends from New York gave their lives in that rescue effort. If our prayers weigh a bit more heavily than those from other states, it's because we came to know so many New York firefighters and police officers in 1995. The New York USAR team was one of the first FEMA groups deployed to Oklahoma City after the Murrah Building bombing, and they were with us through some of the most difficult days and nights of the rescue and recovery effort–a period much like New York experienced after September 11, with mixed hope and despair.

September 11, 2001

"People engrossed by the tragedies in New York, Washington, and Pennsylvania are looking for heroes and are finding them in firefighters. These are people who have a real commitment to saving lives, and we feel so threatened and so vulnerable that we're now sort of perceiving them in a way that we hadn't quite in the past." Jill Stein, sociologist, University of California at Los Angeles

Photo courtesy of FEMA

Those feelings came back to us in Oklahoma after the New York attack. Some of those who survived the 1995 bombing are having an especially difficult time with the events in New York, Washington, and Pennsylvania. It's a nightmare replayed, on a much larger scale, and our wounds are still raw. But our grief is tempered by memories of the good men and women who stood with us six years ago. On that tragic September 11, as I watched the endless replays of those tall towers collapsing, I couldn't help but think of Ray Downey. Somehow, I knew he was there.

Ray Downey had been a firefighter for 39 years. He was the most decorated member of the New York Fire Department, perhaps the nation's top expert in collapse rescue. In Oklahoma City, he diagnosed the tottering, rain-soaked Murrah Building with a practiced eye and worked with us to shore up the structure so rescue and recovery operations could proceed.

Ray named the huge concrete slab that hung ominously overhead. "That's 'Mother'," he said simply, as if it had no power to fall on him. He worked with local firefighters and other FEMA teams to place support beams that prevented the rubble from shifting. I have no doubt that he saved lives in Oklahoma City.

FALLEN HEROES

I looked forward to seeing Ray each day at the bomb site, and I hope he felt the same. We bantered about our shared Catholic religion, and when some nuns from Germany sent some rosaries to my office, I went in search of Ray to give one to him. A year later, when I visited New York to pin yet another decoration on Ray Downey and his fellow fire and police heroes, he was still wearing that rosary.

I hope he had it with him on September 11.

When the New York USAR team filed their after-action report on their tour of duty in Oklahoma City, they ended it with four words: "God Bless You All." Today, Oklahoma returns that sentiment to New York, and especially to all those members of the police and fire service who died, were injured, or who remain missing.

Not long before he died, Ray Downey attended the funeral of another New York firefighter who had given his life in the line of duty.

"Sometimes," he told a reporter, "goodbye is really goodbye."

I refuse to believe that, Ray. You, and those who so proudly wore the badge and principles for which you died, will always be with us in Oklahoma.

May God Bless You All.

Photo courtesy of Fire Engineering

SEPTEMBER 11, 2001

(From Left) Matt Rush, Fireman, Austin (TX) FD; Mark Wesseldine, Fireman, FDNY; Tom Murray, Captain, San Francisco (CA) FD; Rick Fritz, Captain, High Point (NC) FD; Andy Fredericks, FDNY; Paul Schuller, Captain, San Jose (CA) FD

OUR BROTHER, ANDREW FREDERICKS

By Fireman Jay Comella, Oakland (CA) Fire Dept.
and Battalion Chief Ted Corporandy,
San Francisco (CA) Fire Dept.

ON SEPTEMBER 11, 2001 THE FIRE SERVICE LOST HUNDREDS OF BROTHERS. AMONG THE RANKS WERE SOME OF THE MOST TALENTED FIREFIGHTERS AND GREATEST LEADERS OUR PROFESSION HAS EVER PRODUCED.

Fireman Andy Fredericks was one such man. He was inspirational, dedicated, and brilliant. The consummate engineman, his utmost passion was for the basic tools of our trade: Hose and nozzle. His greatest concern was the most efficient use of these fundamental weapons of firefighting. At the heart of his message was the superiority of the smoothbore nozzle over the fog nozzle. His research, study, writing, and subsequent teaching on this most basic fireground element (the nucleus around which all other operations revolve) was incredibly detailed and intricate.

Andy left us with a wealth of knowledge, but it would be naïve to think any one member of the fire service could fill the void created by his loss. He touched many of us across the country through his articles, hands-on training, lectures, and videos. He created many disciples of his no-nonsense, back-to-basics approach to combating today's ever more complicated and dangerous fire problems. Those fortunate firefighters who benefited from his instruction must carry on his work with renewed commitment. President Lincoln stated in the Gettysburg Address, "…It is for us, the living, rather to be dedicated here to the unfinished work which, they have, thus far, so nobly carried on." These words are fitting. Only collectively can we have the same positive effect on the Fire Service that Andy had as an individual.

Andy's passion for his work as an author, educator, instructor, and trainer resulted in improvements in fireground operational efficiency and safety. These improvements have no doubt saved the lives of civilians as well as firefighters. This is his legacy to all of us. For those of us in the Fire Service he is a shining example of dedication, character, professionalism, and love for mankind.

We have shed many tears over the loss of our brother, friend, and mentor, but this is tempered by the memory of his unrelenting humor and his quick and sarcastic wit. He always kept us on our toes and he always had us laughing. Over the years tears may diminish, but our memory will never fade.

To our brothers in FDNY and Squad 18: The entire Fire Service shares the burden of your grief. To Andy's family (wife Michelle and children Andrew and Haley) to whom he was so devoted and loved so much: Our thoughts and prayers will always be with you.

God bless you Andy. Life is not the same without you.

Photo by Mark Seliger

SEPTEMBER 11, 2001

Photo by Time, Inc.

ALL THAT MY CHILDREN HAVE TAUGHT ME

by Capt. Michael Veseling
Engine Co. #3, City of Naperville (IL) Fire Dept.

ON THE AFTERNOON OF SEPTEMBER 11TH, MY YOUNG DAUGHTER CAME HOME FROM 1ST GRADE AND GAVE ME A BIG HUG SAYING, " I AM SORRY SO MANY OF YOUR FRIENDS DIED DADDY, I LOVE YOU. AND DON'T WORRY DADDY, THEY WILL BE OK SOON." NEXT IT WAS MY SON'S TURN. HE GAVE ME A BIG HUG, AND ASKED THAT WE PLAY TOGETHER.

41

Fallen Heroes

We sat on the family room floor, took out his 30 Fire Department Hot Wheels vehicles…engines, trucks, rescues and chief's cars, and we played fireman. We "transmitted the box," we forced entry, we squirted water, and we struck out the alarm. And after a while, I asked him, "What about saving the people? Don't we have to rescue anyone?" With that, his 3-year-old little eyes looked to me and he calmly said, "Of course Daddy, they saved the people, and now it's time for the firemen to go to sleep."

And as we continued to play, as my daughter made frequent trips to hug me and tell me that she loves me, not to mention to ask for a snack before bed, these words and actions of my little ones began to make sense. I thought of those "friends who died… and will be okay soon", of Mike Esposito, Tim Stackpole, Andy Fredericks, and so many more, and I thought… " they have saved the people, and now it's time to go to sleep."

As our Brothers rest in peace, we here are left only with feelings of sorrow. We mourn over the loss of their leadership, their humor, their ability to overcome all odds, and their courage to engage the mightiest of enemies. We grieve at our own losses and for the voids left in our hearts. Yet as we grieve, we must also remember that we have already received these blessings and gifts from our noble Brothers. Their leadership, engraved forever in our minds, has become the benchmark of the American Fire Service. Their humor, displayed in all they have done, serves to comfort us in times of trial and doubt. Their courage, in the face of a mighty foe, and their ability to overcome all odds, displayed for all to see… defining for us, our children, and our children's children, all that is a hero.

A month and several days have now passed, and as I sit alone, my children nestled in bed, I look on the family room wall, at photographs of these heroes. One such picture, of a cold January morning at Squad 1, holds my attention for a moment. This picture, of Brothers gathered around the squad, and Mike Esposito beside me in his red plaid "Elmer Fudd" hat, reminds me of the gifts I have received. It reminds me that I have learned, that I have laughed, and that I will forever be indebted to men like Mike Esposito. Men who challenge and inspire us, men who have taught, have led, and who in their death, have shown a nation how to live.

May God's blessings surround the entire Esposito Family, and all the friends of our fallen Brother Michael.

Photos by Time, Inc.

SEPTEMBER 11, 2001

Photo by Time, Inc.

THE HEART & SOUL OF SQUAD 18

By John Ceriello, Squad 18, FDNY

DAYS AFTER SEPTEMBER 11TH, I WAS HAVING A CONVERSATION WITH BOB STUDE, A PAST 20+ YEAR MEMBER OF ENGINE 18 (NOW SQUAD 18). HE WAS OVERWHELMED WITH GRIEF FOR WHAT HAD JUST HAPPENED. HE WAS RELIVING SOME OF THE EVENTS HE EXPERIENCED ON OCTOBER 17, 1966, WHICH UP UNTIL NOW, WAS THE WORST DAY FOR THE FDNY: TWELVE MEN CAUGHT IN A COLLAPSE ON 23RD STREET.

43

Fallen Heroes

Photo by Richard Smulczeski/FdnyPhotography.com

Photo by Time, Inc.

He thought that there would never be a day like that again. We now know, this was not meant to be.

When I think of what Squad 18 lost, I am reminded of the vast differences in the men who came together in early 1998 to transform Engine 18 to Squad 18. Leading the way was Lt. Billy McGinn, the heart and soul of the company. He always put his men first, was eager to work, and often found himself and his company in the thick of things. He knew our procedures well and would often school us on them and show guidance when needed. His aggressiveness was tempered with intelligence–a great combination for a leader. When I heard we had what looked like the biggest fire FDNY would ever face, we all knew Billy would have to be on duty.

Eric Allen came from Engine 217 Bed-Stuy Brooklyn via Squad 288 in Queens. He never lost his true roots as an Engine man. Small in stature, huge in heart ."Stay low and go." He had no choice with staying low, but go he did. Eric was our resident Jack-of-All-Trades. The beginning and end of each tour, Eric would carry all of his 5 gallon buckets in and out of the Firehouse on West 10th Street. We kidded him that he had broad shoulders due to this personal exercise program of his. He was never happier than when he was talking about his wife Kiki, and the birth of their daughter three years ago. Always helping out someone no matter what needed to be done, Eric was a true friend to all.

September 11, 2001

Andrew Fredericks. What can I say? Countless firefighters knew him well, either personally or through his writings and lecture series. He came to us from Engine 48 in The Bronx, but I think he never totally left there. He often would tell us fire stories from "Da Bronx" and our sarcastic reply would be "Why don't you go back there?" (Of course we did not mean it.) Andy raised our bar at Squad 18. His extensive knowledge transcended to a plane where many of us did not go. For me, Andy was the source of the fringe elements of the Fire Service. When I needed information on subjects such as Fire Protection or response patterns, I called Andy. He was a wealth of information. His insight into the psychological aspect of firefighters was something he was just starting to reveal in his writings.

Dave Halderman. A quiet man with a smile. Dave's journey brought him to us through Engine 209 in Brooklyn. Dave did not pound his chest like some. He was reserved, going about his business with his own resolve. I recently saw that he had become confident in his knowledge of all the technical rescue aspects of Squad work. It showed in his face. Dave was fun to have around the Firehouse, because he always took the ribbing well. I would comment in the morning that his hair looked like he had combed it with a pack of firecrackers. He would just smile and agree. He was a good man.

"We have met the worst of humanity with the best of humanity. Our skyline will rise again." Mayor Rudolph Giuliani

Photo by Fire Engineering

Timmy Haskell. Part model, part carpenter, part boat captain but all Fireman. He worked in TriBeca, Ladder 8 before Squad 18. Timmy went at 100 miles an hour: Getting to the roof to cut a trench on an "H" type building (then sticking his arm and half his torso into the trench to knock down the ceiling, almost barbequing himself). Running out of the Firehouse to meet an engineer to design something for his girlfriend's showroom uptown. He was

Fallen Heroes

non-stop, never slow down. The morning of the 11th, Timmy had already been relieved of duty when the hell-bound hi-jackers on the 767 flew right over the Firehouse. Timmy did not hesitate driving Squad 18's Haz-Mat rig down to the World Trade Center.

Manny Mojica was your stereotypical, muscle-bound Harley man – on the outside. Inside, he was one of the kindest men you could ever know. Manny was an original Engine 18 member. When we all arrived in Manhattan's West Village, he showed us the neighborhood, from the most dangerous buildings to the best watering holes. Manny knew them all well. In my time working with him, I don't think I ever heard him say a bad word about anybody or anything. It just wasn't his style. But you always knew when Manny was coming to work: You could hear his Harley roaring up W. 10th Street, always coming in the wrong way on a one-way street.

Larry Virgilio, last but certainly not least. Larry was an enigma. He was so multi-faceted that I am sure I will not do him justice. Before coming to Squad 18, Larry worked on the upper west side of Manhattan in Ladder Co. 35, located underneath Lincoln Center, which is a Cultural Arts Center. He loved the Arts, but most of all he loved being a fireman, excelling in the technical aspects of the job. Ropes were his specialty. We would go rock climbing just north of the city (yes, the concrete jungle of New York ends just a short distance outside the city). He would be our lead climber showing us our route up the 200 ft. face of the cliff. Larry also made wine–good wine. He had his own label designed by Lt. Pete Campanelli, called Tenement Cellars. We all helped him on his grape crushing days and it was great fun. After the 11th, we bottled his last batch. His girlfriend Abigail wanted us to be there. It was a somber remembering.

We will miss each and everyone of our Fallen Brothers. We will never forget them. I hope that I have given a little insight into these special men. Although this catastrophe was localized in the NY tri-state area, it was also a blow to our Nation and our American Fire Service. For as we all know, when an attack occurs on our soil, the frontline soldiers will be the firefighters who will shoulder the brunt of the assault. For me the thought of the dedication and sacrifice that my brothers showed in their final heroic acts will carry me on to help rebuild this department and not let them be forgotten.

"We are shaken, but we are not defeated. We stare adversity in the eye, and we move on."
Fire Commissioner Thomas Von Essen

Photo by Time, Inc.

SEPTEMBER 11, 2001

THE WORLD REACTS

By Geoff Williams, Firefighter
Scotland, UK

ON BEHALF OF ALL FIREFIGHTERS AND STAFF OF THE CENTRAL SCOTLAND FIRE BRIGADE, I WOULD LIKE TO EXPRESS OUR DEEPEST SYMPATHY TO ALL OUR BROTHERS, THEIR FAMILIES, AND CITIZENS OF THE UNITED STATES WHO HAVE BEEN INVOLVED WITH THESE HORRENDOUS ATROCITIES. WORDS FAIL TO EXPRESS THE ANGER AND SADNESS THAT I PERSONALLY FEEL INSIDE, HAVING LOST SO MANY REAL BROTHERS FROM ACROSS THE POND.

Fallen Heroes

Ray Downey dead? I still cannot believe or accept that he is not with us. Our kinfolk have been kicked bloody hard, and over here your fellow firefighter brothers are indeed limping.

I read an article recently that said there is an old Tibetan proverb, "Goodness speaks in a whisper. Evil shouts." On September 11, 2001 evil did indeed shout in your great city. It shouted loudly enough to be heard all around the world, and its echoes are still being heard. However just beneath that roar, if you listen carefully, you will hear "Goodness" whispers. Whispers in the bravery of the FDNY and Washington fire fighters and the Urban Search and Rescue teams and the people in New York and Washington. These are the people that have stuck to their task quietly and let everyone else consider the consequences of this heinous crime.

Having seen my home city, Manchester, blasted to bits over many, many years by terrorism, I know a wee, small bit of what you feel about the physical damage your great country feels. I suppose "thankfully," our dead have always been in the low figures of eights and nines...never on the scale we have all just witnessed, but a life is a life, and it's a family that is crushed, a dad or baby who is no more.

Over here in my brigade, my firefighters immediately said that they wanted to go and help. Needless to say there was no

"The appalling terrorist attack on the USA, and in particular on New York itself, shows abhorrent cowardice and unbelievable contempt for human life. The reaction everywhere has been one of shock, utter bewilderment, and pain. I am most concerned to convey to you – in the name of the entire Frankfurt Book Fair team and without doubt, the world-wide Frankfurt community – our deepest sympathy and unqualified solidarity."
Lorenzo A. Rudolf, Director of the Franfurt Book Fair, Germany

"Please accept our warmest condolences for the tragedy that has fallen on your country. Now in particular we feel how important and close to our hearts are distant Americans. We cross our fingers for you and wish you moral and physical strength in recovering from this world disaster." Vladimir Stabnikov, Olympus Business Publishers, Russia

way they could, so they did the next best thing and started to collect for their brothers' families.

They took the fire trucks into the town of Stirling (where I live), and placed a large Stars and Stripes flag over the truck and started to collect money. Within a very short time they had raised £20,000–which is US$ 25,000! Then they had a fire call. Once they re-commenced, they hit it again...to cut a long story short, they raised approximately US$45,000 (forty thousand pounds) for their U.S. kinfolk in just over 12 hours!

This is unbelievable for the size of Stirling Town; everyone and his dog wanted to contribute. When we normally have a Fire Service Benevolent Collection, we usually make around US$15,000–if we are lucky. Which is still very good, but this had to be seen to believed. The guys were mobbed. Many American tourists who were visiting the old castle came and said that they couldn't believe what their eyes were telling them. Everyone was rushing to add their bit. So although "Evil shouts," the Goodness whispers hit a high note those two days...in Stirling. I want you to please pass on this small message to let the guys and gals know that we really do care and are so proud of you...words cannot do you all justice.

To conclude (having just returned from Texas where I attended an FETN Board meeting), I have never felt so sad and inadequate as I did at that meeting trying to explain how we Brits are so angered at this atrocity towards our brothers. Perhaps, God willing, we'll still manage to get together at the FDIC in Indianapolis; although at this moment in time my mind can't see anything but funerals.

Take care, mates–our thoughts and prayers are with you all.

FALLEN HEROES

"We are completely shocked on hearing of the attacks. We share the grief of your countrymen in this grave tragedy and pray for the safety of your family, friends, and colleagues."

Aditya Prakash, India

THE FALLEN HEROES OF SEPTEMBER 11, 2001

"Just a phenomenal group of people. The best of the department."

FDNY Fire Commissioner Thomas Von Essen

FALLEN HEROES

For Those Firefighters Who Retired Yet Returned To Give Their All.

They had fulfilled their duties to the department.

There was no requirement that they show up at the scene–except for their inner need to serve their fellow man and to help their brothers.

James J. Corrigan
Retired Captain
Engine 320
Memorial Service: October 13, 2001

Philip T. Hayes
Retired Firefighter
Engine 217
Memorial Service: October 20, 2001

William Wren
Retired Firefighter
Ladder 166
Memorial Service: November 3, 2001

Joseph Agnello
Firefighter
Ladder 118
Memorial Service: October 10, 2001

Brian Ahearn
Lieutenant
Engine 230
Memorial Service: October 27, 2001

September 11, 2001

Eric Allen

Firefighter

Squad 18

Laid to Rest: September 20, 2001

Richard Allen

Firefighter

Ladder 15

Memorial Service: November 9, 2001

James Amato

Captain

Squad 41

Laid to Rest: September 28, 2001

Calixto Anaya, Jr.

Firefighter

Engine 4

Memorial Service: November 8, 2001

FALLEN HEROES

Joseph Angelini, Jr.

Firefighter

Ladder 4

Memorial Service: October 22, 2001

Joseph Angelini

Firefighter

Rescue 1

Laid to Rest: September 22, 2001

Faustino Apostol, Jr.

Firefighter

Battalion 2

Laid to Rest: October 6, 2001

David Arce

Firefighter

Engine 33

Memorial Service: November 5, 2001

SEPTEMBER 11, 2001

Louis Arena

Firefighter

Ladder 5

Laid to Rest: September 19, 2001

Carl F. Asaro

Firefighter

Battalion 9

Memorial Service: October 27, 2001

Gregg Atlas

Lieutenant

Engine 10

Memorial Service: October 29, 2001

Gerald Atwood

Firefighter

Ladder 21

Missing

FALLEN HEROES

Gerard Baptiste

Firefighter

Ladder 9

Memorial Service: November 16, 2001

Gerard Barbara

Assistant Chief

Citywide Tour Commander

Memorial Service: October 1, 2001

Matthew Barnes

Firefighter

Ladder 25

Memorial Service: November 9, 2001

Arthur Barry

Firefighter

Ladder 15

Memorial Service: October 13, 2001

September 11, 2001

Steven Bates

Lieutenant

Engine 235

Memorial Service: October 6, 2001

Carl Bedigian

Lieutenant

Engine 214

Laid to Rest: November 5, 2001

Stephen Belson

Firefighter

Ladder 24

Memorial Service: September 29, 2001

John Bergin

Firefighter

Rescue 5

Memorial Service: October 20, 2001

FALLEN HEROES

Paul Beyer

Firefighter

Engine 6

Missing

Peter Bielfeld

Firefighter

Ladder 42

Memorial Service: November 3, 2001

Brian Bilcher

Firefighter

Squad 1

Missing

Carl Bini

Firefighter

Rescue 5

Memorial Service: November 3, 2001

SEPTEMBER 11, 2001

Christopher Blackwell

Firefighter

Rescue 3

Memorial Service: October 20, 2001

FIRE ENGINEERING INSTRUCTOR

Blackwell served in the special Collapse Rescue unit for the last 12 years. He received several commendations for bravery, including a medal for helping remove people from an airplane that had slipped off the runway at LaGuardia Airport. He was a teacher for the State of New York and for the Fire Department Instructors Conference.

Michael Bocchino

Firefighter

Battalion 48 (Working in Engine Co. 239)

Memorial Service: October 13, 2001

Frank Bonomo

Firefighter

Engine 230

Memorial Service: October 2, 2001

FALLEN HEROES

Gary Box

Firefighter

Squad 1

Memorial Service: October 10, 2001

Michael Boyle

Firefighter

Engine 33

Memorial Service: November 3, 2001

Kevin Bracken

Firefighter

Engine 40

Memorial Service: November 7, 2001

Michael Brennan

Firefighter

Ladder 4

Memorial Service: October 12, 2001

September 11, 2001

Peter Brennan

Firefighter

Rescue 4

Memorial Service: October 1, 2001

Patrick Brown

Captain

Ladder 3

Memorial Service: November 9, 2001

FIRE ENGINEERING AUTHOR

Captain Brown became well known for his May 1991 dramatic rope rescue of two victims from the window ledge of the 12th floor of a Manhattan high-rise, where the flames were at the victims' backs and there was nothing to tie a rescue rope to.

Daniel Brethel

Captain

Ladder 24

Laid to Rest: September 17, 2001

FALLEN HEROES

Andrew Brunn

Probationary Firefighter

Ladder 5

Laid to Rest: September 18, 2001

Vincent Brunton

Captain

Ladder 105

Missing

Ronald Bucca

Fire Marshal

Laid to Rest: November 10, 2001

Greg Buck

Firefighter

Engine 201

Missing

September 11, 2001

William Burke, Jr.

Captain

Engine 21

Memorial Service: October 25, 2001

Donald Burns

Assistant Chief

Citywide Tour Commander

Memorial Service: October 9, 2001

FIRE ENGINEERING AUTHOR

Chief Burns worked his way up through the ranks. As a captain in research and development, he pushed people to attend the Fire Department Instructors Conference. He taught high-rise firefighting at the New York State Fire Academy. He helped write Fire Engineering's coverage of the World Trade Center Bombing in 1993, which became the official report of the U.S. Fire Administration.

John Burnside

Firefighter

Ladder 20

Memorial Service: November 16, 2001

FALLEN HEROES

Thomas Butler

Firefighter

Squad 1

Missing

Patrick Byrne

Firefighter

Ladder 101

Memorial Service: November 29, 2001

George Cain

Firefighter

Ladder 7

Memorial Service: November 2, 2001

Salvatore Calabro

Firefighter

Ladder 101

Memorial Service: November 19, 2001

SEPTEMBER 11, 2001

Frank Callahan

Captain

Ladder 35

Missing

Michael Cammarata

Firefighter

Ladder 11

Memorial Service: October 5, 2001

Brian Cannizzaro

Firefighter

Ladder 101

Laid to Rest: November 15, 2001

Dennis Carey

Firefighter

Haz-Mat Company 1

Laid to Rest: October 9, 2001

FALLEN HEROES

Michael Carlo

Firefighter

Engine 230

Memorial Service: November 30, 2001

Michael Carroll

Firefighter

Ladder 3

Memorial Service: October 20, 2001

Peter Carroll

Firefighter

Squad 1

Laid to Rest: September 17, 2001

Thomas Casoria

Firefighter

Engine 22

Missing

September 11, 2001

Michael Cawley

Firefighter

Ladder 136

Memorial Service: October 7, 2001

Vernon Cherry

Firefighter

Ladder 118

Memorial Service: October 19, 2001

Nicholas Chiofalo

Firefighter

Engine 235

Memorial Service: November 3, 2001

John Chipura

Firefighter

Engine 219

Memorial Service: November 6, 2001

FALLEN HEROES

Michael Clarke

Firefighter

Ladder 2

Laid to Rest: October 10, 2001

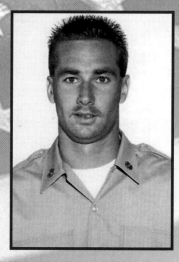

Steven Coakley

Firefighter

Engine 217

Memorial Service: November 3, 2001

Tarel Coleman

Firefighter

Squad 252

Memorial Service: November 17, 2001

John Collins

Firefighter

Ladder 25

Memorial Service: September 29, 2001

SEPTEMBER 11, 2001

Robert Cordice

Firefighter

Engine 152 (Working in Squad 1)

Memorial Service: November 9, 2001

Ruben Correa

Firefighter

Engine 74

Memorial Service: October 4, 2001

James Coyle

Firefighter

Ladder 3

Memorial Service: October 24, 2001

Robert Crawford

Firefighter

Safety Battalion 1

Memorial Service: November 10, 2001

FALLEN HEROES

John Crisci

Lieutenant

Haz-Mat Company 1

Laid to Rest: September 28, 2001

Dennis Cross

Battalion Chief

Battalion 57

Laid to Rest: September 22, 2001

Thomas Cullen III

Firefighter

Squad 41

Memorial Service: October 5, 2001

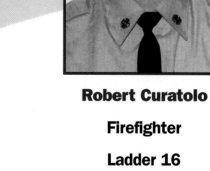

Robert Curatolo

Firefighter

Ladder 16

Laid to Rest: September 18, 2001

Edward D'Atri

Lieutenant

Squad 1

Memorial Service: October 12, 2001

Michael D'Auria

Firefighter

Engine 40 (Probationary Class 2 of 2001)

Memorial Service: October 20, 2001

Scott Davidson

Firefighter

Ladder 118

Memorial Service: October 13, 2001

Edward Day

Firefighter

Ladder 11

Memorial Service: October 6, 2001

FALLEN HEROES

Thomas DeAngelis

Battalion Chief

Battalion 8

Memorial Service: October 13, 2001

Manuel Delvalle

Firefighter

Engine 5

Laid to Rest: October 13, 2001

Martin DeMeo

Firefighter

Haz-Mat Company 1

Laid to Rest: October 9, 2001

David DeRubbio

Firefighter

Engine 226

Memorial Service: November 10, 2001

SEPTEMBER 11, 2001

Andrew Desperito

Lieutenant

Engine 1

Laid to Rest: September 18, 2001

Dennis Devlin

Battalion Chief

Battalion 9

Memorial Service: September 29, 2001

Gerard Dewan

Firefighter

Ladder 3

Memorial Service: October 27, 2001

George DiPasquale

Firefighter

Ladder 2

Memorial Service: October 8, 2001

FALLEN HEROES

Kevin Donnelly

Lieutenant

Ladder 3

Memorial Service: October 6, 2001

Raymond Downey

Battalion Chief

Special Operations Command

Memorial Service: December 15, 2001

FIRE ENGINEERING AUTHOR

Chief Downey was the USAR task force leaders' representative to FEMA for all 27 teams, and a member of FEMA's Advisory Committee. He commanded rescue operations at the Oklahoma City Bombing and served on numerous terrorism task forces and panels. He was an advisory board member of Fire Engineering and of the Fire Department Instructors Conference, and was a teacher to firefighters nationwide through H.O.T. classes at FDIC and FDIC West.

Kevin Dowdell

Lieutenant

Rescue 4

Missing

SEPTEMBER 11, 2001

Gerard Duffy

Firefighter

Ladder 21

Memorial Service: October 6, 2001

Martin Egan Jr.

Captain

Division 15

Laid to Rest: September 20, 2001

Michael Elferis

Firefighter

Engine 22

Laid to Rest: November 27, 2001

Francis Esposito

Firefighter

Engine 235 (Assigned to Ladder 79)

Memorial Service: November 2, 2001

FALLEN HEROES

Michael Esposito

Lieutenant

Squad 1

Memorial Service: November 16, 2001

Robert Evans

Firefighter

Engine 33

Memorial Service: November 28, 2001

John Fanning

Battalion Chief

Haz-Mat Operations

Memorial Service: October 6, 2001

Thomas Farino

Captain

Engine 26

Memorial Service: November 8, 2001

September 11, 2001

Terrence Farrell

Firefighter

Rescue 4

Laid to Rest: November 1, 2001

Joseph Farrelly

Captain

Engine 4

Memorial Service: October 15, 2001

William Feehan

First Deputy Commissioner

Formerly 23rd Chief of Department

Laid to Rest: September 15, 2001

Lee Fehling

Firefighter

Engine 235

Memorial Service: October 9, 2001

FALLEN HEROES

Alan Feinberg

Firefighter

Battalion 9

Missing

Michael Fiore

Firefighter

Rescue 5

Memorial Service: October 27, 2001

John Fischer

Captain

Ladder 20

Memorial Service: November 8, 2001

Andre Fletcher

Firefighter

Rescue 5

Memorial Service: October 13, 2001

September 11, 2001

John Florio

Firefighter

Engine 214

Laid to Rest: November 3, 2001

Michael Fodor

Lieutenant

Ladder 21

Memorial Service: October 27, 2001

Thomas Foley

Firefighter

Rescue 3

Laid to Rest: September 29, 2001

David Fontana

Firefighter

Squad 1

Memorial Service: October 17, 2001

FALLEN HEROES

Robert Foti

Firefighter

Ladder 7

Memorial Service: October 6, 2001

Andrew Fredericks

Firefighter

Squad 18

Laid to Rest: October 8, 2001

FIRE ENGINEERING AUTHOR

Fredericks was a New York State-certified fire instructor at the Rockland County Fire Training Center in Pomona, New York, and an adjunct instructor at the New York State Academy of Fire Science. He held a master's degree in fire protection management, and taught at the University of Illinois Fire Service Institute. Fredericks developed the Fire Engineering "Bread and Butter" videos and was a H.O.T. class instructor at the Fire Department Instructors Conference.

Peter Freund

Lieutenant

Engine 55

Laid to Rest: October 30, 2001

September 11, 2001

Thomas Gambino, Jr.

Firefighter

Rescue 3

Memorial Service: October 15, 2001

Peter J. Ganci, Jr.

28th Chief of Department

Laid to Rest: September 15, 2001

Charles Garbarini

Lieutenant

Battalion 9

Memorial Service: October 6, 2001

Thomas Gardner

Firefighter

Haz-Mat Co. 1

Memorial Service: October 19, 2001

Fallen Heroes

Matthew Garvey

Firefighter

Squad 1

Laid to Rest: October 30, 2001

Bruce Gary

Firefighter

Engine 40

Laid to Rest: October 6, 2001

Gary Geidel

Firefighter

Rescue 1

Missing

Edward Geraghty

Battalion Chief

Battalion 9

Memorial Service: October 25, 2001

September 11, 2001

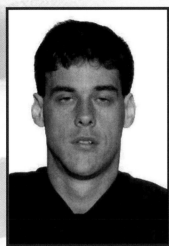

Denis Germain

Firefighter

Ladder 2

Memorial Service: September 29, 2001

Vincent Giammona

Lieutenant

Ladder 5

Memorial Service: October 3, 2001

James Giberson

Firefighter

Ladder 35

Memorial Service: October 5, 2001

Ronnie Gies

Firefighter

Squad 288

Memorial Service: October 5, 2001

Fallen Heroes

Paul Gill

Firefighter

Engine 54

Memorial Service: October 20, 2001

John Ginley

Lieutenant

Engine 40

Memorial Service: October 22, 2001

Jeffrey Giordano

Firefighter

Ladder 3

Memorial Service: October 13, 2001

John Giordano

Firefighter

Engine 37

Memorial Service: October 2, 2001

September 11, 2001

Keith Glascoe

Firefighter

Ladder 21

Memorial Service: November 17, 2001

James Gray

Firefighter

Ladder 20

Memorial Service: October 27, 2001

Joseph Grzelak

Battalion Chief

Battalion 48

Memorial Service: November 17, 2001

Jose Guadalupe

Firefighter

Engine 54

Laid to Rest: October 1, 2001

FALLEN HEROES

Geoffrey Guja

Lieutenant

Battalion 43

Laid to Rest: September 22, 2001

Joseph Gullickson

Lieutenant

Ladder 101

Memorial Service: November 10, 2001

David Halderman

Firefighter

Squad 18

Laid to Rest: October 1, 2001

Vincent Halloran

Lieutenant

Ladder 8

Memorial Service: November 9, 2001

SEPTEMBER 11, 2001

Robert Hamilton

Firefighter

Squad 41

Memorial Service: September 27, 2001

Dana Hannon

Firefighter

Engine 26

Memorial Service: December 8, 2001

FIRE ENGINEERING INSTRUCTOR

Hannon began his fire service career as a volunteer in the Wyckoff (NJ) Fire Department, where he attained the rank of captain. He was hired by the Bridgeport (CT) Fire Department where he earned a Medal of Valor for the rescue of an elderly civilian. He was a Hands-On Training instructor at the Fire Department Instructors Conference.

Sean Hanley

Firefighter

Ladder 20

Laid to Rest: September 18, 2001

FALLEN HEROES

Thomas Hannafin

Firefighter

Ladder 5

Laid to Rest: September 20, 2001

Daniel Harlin

Firefighter

Ladder 2

Missing

Harvey Harrell

Lieutenant

Rescue 5

Memorial Service: October 12, 2001

Stephen Harrell

Lieutenant

Battalion 7 (Covering in Ladder 10)

Memorial Service: October 10, 2001

September 11, 2001

Timothy Haskell

Firefighter

Squad 18

Laid to Rest: September 19, 2001

Terence Hatton

Captain

Rescue 1

Laid to Rest: October 4, 2001

FIRE ENGINEERING AUTHOR

Captain Hatton previously was a firefighter in Rescue 2 and a lieutenant in Rescue 4. He realized the need for a technical rescue school and helped FDNY create one of the most disciplined fire department rescue schools in the country.

Thomas Haskell, Jr.

Battalion Chief

Ladder 132 (Assigned to Division 15)

Memorial Service: November 10, 2001

Fallen Heroes

Michael Haub

Firefighter

Ladder 4

Memorial Service: October 10, 2001

Michael Healey

Lieutenant

Squad 41

Memorial Service: November 11, 2001

John Heffernan

Firefighter

Ladder 11

Laid to Rest: October 4, 2001

Ronnie Henderson

Firefighter

Engine 279

Memorial Service: October 14, 2001

SEPTEMBER 11, 2001

Joseph Henry

Firefighter

Ladder 21

Memorial Service: October 5, 2001

William Henry

Firefighter

Rescue 1

Laid to Rest: September 20, 2001

Thomas Hetzel

Firefighter

Ladder 13

Laid to Rest: October 9, 2001

Brian Hickey

Captain

Rescue 4

Missing

FALLEN HEROES

Timothy Higgins

Lieutenant

Special Operations

Laid to Rest: October 4, 2001

Jonathan Hohmann

Firefighter

Haz-Mat Co. 1

Memorial Service: October 7, 2001

Thomas Holohan

Firefighter

Engine 6

Laid to Rest: September 28, 2001

Joseph Hunter

Firefighter

Squad 288

Memorial Service: November 10, 2001

SEPTEMBER 11, 2001

Walter Hynes

Captain

Ladder 13

Laid to Rest: September 19, 2001

Jonathan Ielpi

Firefighter

Squad 288

Memorial Service: October 27, 2001

Frederick III, Jr.

Captain

Ladder 2

Laid to Rest: October 9, 2001

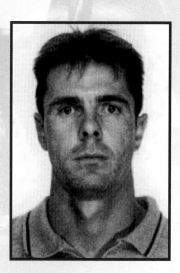

William Johnston

Firefighter

Engine 6

Laid to Rest: October 5, 2001

FALLEN HEROES

Andrew Jordan

Firefighter

Ladder 132

Missing

Karl Joseph

Firefighter

Engine 207

Memorial Service: November 17, 2001

Anthony Jovic

Lieutenant

Ladder 34

Working in Engine 279

(Formerly Firefighter, Ladder 150)

Memorial Service: November 12, 2001

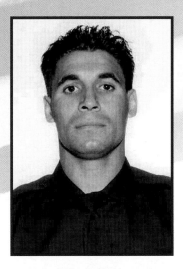

Angel Juarbe, Jr.

Firefighter

Ladder 12

Memorial Service: October 27, 2001

SEPTEMBER 11, 2001

Mychal Judge

Chaplain

Laid to Rest: September 15, 2001

Vincent Kane

Firefighter

Engine 22

Memorial Service: October 6, 2001

Charles Kasper

Battalion Chief

Special Operations

Memorial Service: September 28, 2001

Paul Keating

Firefighter

Ladder 5

Memorial Service: October 5, 2001

FALLEN HEROES

Richard Kelly, Jr.

Firefighter

Ladder 11

Memorial Service: October 3, 2001

Thomas Kelly

Firefighter

Ladder 105

Memorial Service: September 29, 2001

Thomas Kelly

Firefighter

Ladder 15

Memorial Service: October 2, 2001

Thomas Kennedy

Firefighter

Ladder 101

Memorial Service: October 24, 2001

September 11, 2001

Ronald Kerwin

Lieutenant

Squad 288

Memorial Service: December 1, 2001

Michael Kiefer

Firefighter

Ladder 132

Missing

Robert King, Jr.

Firefighter

Engine 33

Missing

Scott Kopytko

Firefighter

Ladder 15

Memorial Service: December 2, 2001

FALLEN HEROES

William Krukowski

Firefighter

Ladder 21

Memorial Service: September 29, 2001

Kenneth Kumpel

Firefighter

Ladder 25

Memorial Service: October 13, 2001

Thomas Kuveikis

Firefighter

Squad 252

Missing

David LaForge

Firefighter

Ladder 20

Memorial Service: October 5, 2001

SEPTEMBER 11, 2001

William Lake

Firefighter

Rescue 2

Laid to Rest: October 13, 2001

Robert Lane

Firefighter

Engine 55

Missing

Peter Langone

Firefighter

Squad 252

Memorial Service: October 26, 2001

Scott Larsen

Firefighter

Ladder 15

Missing

FALLEN HEROES

Joseph Leavey

Lieutenant

Ladder 15

Laid to Rest: November 13, 2001

Neil Leavy

Firefighter

Engine 217

Laid to Rest: October 4, 2001

Daniel Libretti

Firefighter

Rescue 2

Laid to Rest: October 16, 2001

Carlos Lillo

Paramedic

Battalion 49

Memorial Service: November 3, 2001

SEPTEMBER 11, 2001

Robert Linnane

Firefighter

Ladder 20

Memorial Service: October 6, 2001

Michael Lynch

Firefighter

Ladder Company 32

(on Rotation Engine 40)

Celebration Mass: December 7, 2001

Michael Lynch

Firefighter

Ladder 4

Memorial Service: November 24, 2001

Michael Lyons

Firefighter

Squad 41

Memorial Service: October 6, 2001

FALLEN HEROES

Patrick Lyons

Firefighter

Squad 252

Missing

Joseph Maffeo

Firefighter

Ladder 101

Memorial Service: December 3, 2001

William Mahoney

Firefighter

Rescue 4

Laid to Rest: November 5, 2001

Joseph Maloney

Firefighter

Ladder 3

Laid to Rest: September 22, 2001

SEPTEMBER 11, 2001

Joseph Marchbanks, Jr.

Battalion Chief

Battalion 12

Memorial Service: October 26, 2001

Charles Margiotta

Lieutenant

Ladder 85

Memorial Service: November 6, 2001

Kenneth Marino

Firefighter

Rescue 1

Memorial Service: October 16, 2001

John Marshall

Firefighter

Ladder 27 (Detailed to Engine 23)

Memorial Service: November 17, 2001

FALLEN HEROES

Peter Martin

Lieutenant

Rescue 2

Memorial Service: October 20, 2001

Paul Martini

Lieutenant

Engine 201

Memorial Service: October 27, 2001

Joseph Mascali

Firefighter

Rescue 5

Memorial Service: October 10, 2001

Keithroy Maynard

Firefighter

Engine 33

Memorial Service: November 14, 2001

SEPTEMBER 11, 2001

Brian McAleese

Firefighter

Engine 226

Memorial Service: December 7, 2001

John McAvoy

Firefighter

Ladder 3

Memorial Service: September 29, 2001

Thomas McCann

Firefighter

Battalion 8

Missing

William McGovern

Battalion Chief

Battalion 2

Laid to Rest: September 22, 2001

Fallen Heroes

William McGinn

Lieutenant

Squad 18

Laid to Rest: October 5, 2001

FIRE ENGINEERING INSTRUCTOR

Lieutenant McGinn previously was a firefighter in Squad 1. He was known for constantly drilling his squads in high-rises. He was a H.O.T. instructor at the Fire Department Instructors Conference.

Dennis McHugh

Firefighter

Ladder 13

Memorial Service: September 28, 2001

Robert McMahon

Firefighter

Ladder 20

Laid to Rest: November 12, 2001

SEPTEMBER 11, 2001

Robert McPadden

Firefighter

Engine 23

Memorial Service: September 29, 2001

Terence McShane

Firefighter

Engine 308 (On Rotation, Ladder 101)

Laid to Rest: November 17, 2001

Timothy McSweeney

Firefighter

Ladder 3

Memorial Service: October 9, 2001

Martin McWilliams

Firefighter

Engine 22

Laid to Rest: September 18, 2001

FALLEN HEROES

Raymond Meisenheimer

Firefighter

Rescue 3

Memorial Service: October 5, 2001

Charles Mendez

Firefighter

Ladder 7

Memorial Service: October 13, 2001

Steve Mercado

Firefighter

Engine 40

Memorial Service: November 10, 2001

Douglas Miller

Firefighter

Rescue 5

Memorial Service: October 4, 2001

September 11, 2001

Henry Miller, Jr.

Firefighter

Ladder 105

Memorial Service: October 1, 2001

Robert Minara

Firefighter

Ladder 25

Memorial Service: October 17, 2001

Thomas Mingione

Firefighter

Ladder 132

Missing

Paul Mitchell

Lieutenant

Battalion 1

(Formerly Firefighter, Ladder 110)

Memorial Service: November 2, 2001

FALLEN HEROES

Louis Modafferi

Captain

Rescue 5

Memorial Service: October 29, 2001

Dennis Mojica

Lieutenant

Rescue 1

Laid to Rest: September 21, 2001

FIRE ENGINEERING AUTHOR

Lieutenant Mojica had approximately 25 years with the Fire Department, and had used his airplane mechanic's knowledge at various crash scenes in and around New York. He trained the USAR team in Puerto Rico, and was an instructor in FDNY's Technical Rescue School.

Manuel Mojica

Firefighter

Squad 18

Laid to Rest: September 20, 2001

SEPTEMBER 11, 2001

Carl Molinaro

Firefighter

Ladder 2

Laid to Rest: October 12, 2001

Michael Montesi

Firefighter

Rescue 1

Funeral Arrangements Pending

Thomas Moody

Captain

Division 1

Memorial Service: October 27, 2001

John Moran

Battalion Chief

Battalion 49

Memorial Service: September 28, 2001

FALLEN HEROES

Vincent Morello

Firefighter

Ladder 35 (Assigned to E-283)

Memorial Service: November 2, 2001

Christopher Mozzillo

Firefighter

Engine 55 (Assigned to Ladder 148)

Memorial Service: December 3, 2001

Richard Muldowney, Jr.

Firefighter

Ladder 7

Missing

Michael Mullan

Firefighter

Ladder 12

Laid to Rest: October 20, 2001

September 11, 2001

Dennis Mulligan

Firefighter

Ladder 2

Memorial Service: October 6, 2001

Raymond Murphy

Lieutenant

Ladder 16

Laid to Rest: October 5, 2001

Robert Nagel

Lieutenant

Engine 58

Memorial Service: November 17, 2001

John Napolitano

Firefighter

Rescue 2

Memorial Service: October 2, 2001

FALLEN HEROES

Peter Nelson

Firefighter

Rescue 4

Laid to Rest: October 31, 2001

Gerard Nevins

Firefighter

Rescue 1

Memorial Service: October 6, 2001

Dennis Oberg

Firefighter

Ladder 105

Missing

Daniel O'Callaghan

Lieutenant

Ladder 4

Memorial Service: November 10, 2001

SEPTEMBER 11, 2001

Thomas O'Hagan

Lieutenant

Battalion 4

Memorial Service: October 14, 2001

Patrick O'Keefe

Firefighter

Rescue 1

Laid to Rest: September 29, 2001

William O'Keefe

Captain

Division 15

Laid to Rest: September 28, 2001

Kevin O'Rourke

Firefighter

Rescue 2

Laid to Rest: September 28, 2001

FALLEN HEROES

Douglas Oelschlager

Firefighter

Ladder 15

Memorial Service: October 19, 2001

Joseph Ogren

Firefighter

Ladder 3

Memorial Service: September 27, 2001

Samuel Oitice

Firefighter

Ladder 4

Memorial Service: November 3, 2001

Eric Olsen

Firefighter

Ladder 15

Memorial Service: November 17, 2001

September 11, 2001

Jeffrey Olsen

Firefighter

Engine 10

Memorial Service: October 12, 2001

Steven Olson

Firefighter

Ladder 3

Memorial Service: October 16, 2001

Michael Otten

Firefighter

Ladder 35

Memorial Service: November 16, 2001

Jeffrey Palazzo

Firefighter

Rescue 5

Memorial Service: October 9, 2001

FALLEN HEROES

Orio Palmer

Battalion Chief

Battalion Chief 7

Memorial Service: October 13, 2001

Frank Palombo

Firefighter

Ladder 105

Memorial Service: October 20, 2001

Paul Pansini

Firefighter

Engine 10

Memorial Service: October 10, 2001

John Paolillo

Battalion Chief

Battalion 11

Laid to Rest: October 10, 2001

SEPTEMBER 11, 2001

James Pappageorge

Firefighter

Engine 23

Memorial Service: December 14, 2001

Robert Parro

Firefighter

Engine 8

Laid to Rest: October 11, 2001

Durrell "Bronko" Pearsall

Firefighter

Rescue 4

Laid to Rest: November 8, 2001

Glenn Perry

Lieutenant

Battalion 12

Memorial Service: October 15, 2001

FALLEN HEROES

Philip Petti

Lieutenant

Battalion 7

Memorial Service: October 7, 2001

Kevin Pfeifer

Lieutenant

Engine 33

Memorial Service: November 2, 2001

Kenneth Phelan

Lieutenant

Engine 217

Memorial Service: November 11, 2001

Christopher Pickford

Firefighter

Engine 201

Missing

SEPTEMBER 11, 2001

Shawn Powell

Firefighter

Engine 207

Memorial Service: November 10, 2001

Vincent Princiotta

Firefighter

Ladder 7

Memorial Service: November 10, 2001

Kevin Prior

Firefighter

Squad 252

Laid to Rest: October 5, 2001

Richard Prunty

Battalion Chief

Battalion 2

Laid to Rest: September 20, 2001

Fallen Heroes

Lincoln Quappe

Firefighter

Rescue 2

Laid to Rest: September 27, 2001

Michael Quilty

Lieutenant

Ladder 11

Memorial Service: September 29, 2001

Ricardo Quinn

Paramedic

Battalion 57

Memorial Service: September 29, 2001

Leonard Ragaglia

Firefighter

Engine 54

Memorial Service: November 17, 2001

September 11, 2001

Michael Ragusa

Firefighter

Engine 279

Missing

Edward Rall

Firefighter

Rescue 2

Memorial Service: November 14, 2001

Adam Rand

Firefighter

Squad 288

Memorial Service: October 13, 2001

Robert Regan

Lieutenant

Ladder 118

Memorial Service: October 5, 2001

FALLEN HEROES

Donald Regan

Firefighter

Rescue 3

Memorial Service: October 8, 2001

Christian Regenhard

Firefighter

Ladder 131

Memorial Service: October 26, 2001

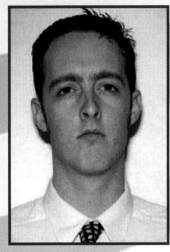

Kevin Reilly

Firefighter

Ladder 40 (Working in Engine 207)

Memorial Service: October 20, 2001

Vernon Richard

Captain

Ladder 7

Memorial Service: December 8, 2001

September 11, 2001

James Riches

Firefighter

Engine 4

Missing

Joseph Rivelli, Jr.

Firefighter

Ladder 25

Memorial Service: November 20, 2001

Michael Roberts

Firefighter

Ladder 35

Memorial Service: November 12, 2001

Michael Roberts

Firefighter

Engine 214

Laid to Rest: October 27, 2001

FALLEN HEROES

Anthony Rodriguez

Firefighter

Engine 279

Memorial Service: October 27, 2001

Matthew Rogan

Firefighter

Ladder 11

Memorial Service: October 2, 2001

Nicholas Rossomando

Firefighter

Rescue 5

Memorial Service: October 30, 2001

Paul Ruback

Firefighter

Ladder 25

Memorial Service: November 3, 2001

September 11, 2001

Stephen Russell

Firefighter

Engine 55

Laid to Rest: November 3, 2001

Michael Russo

Lieutenant

Special Operations

Memorial Service: November 6, 2001

Matthew Ryan

Battalion Chief

Battalion 1

Memorial Service: October 12, 2001

Thomas Sabella

Firefighter

Ladder 13

Memorial Service: October 2, 2001

FALLEN HEROES

Christopher Santora

Firefighter

Engine 54

Memorial Service: December 1, 2001

John Santore

Firefighter

Ladder 5

Laid to Rest: September 19, 2001

Gregory Saucedo

Firefighter

Ladder 5

Memorial Service: September 25, 2001

Dennis Scauso

Firefighter

Haz-Mat Company 1

Memorial Service: November 26, 2001

S EPTEMBER 11, 2001

John Schardt

Firefighter

Engine 201

Memorial Service: November 12, 2001

Fred Scheffold

Battalion Chief

Battalion 12

Memorial Service: October 12, 2001

Thomas Schoales

Firefighter

Engine 4

Memorial Service: September 29, 2001

Gerard Schrang

Firefighter

Rescue 3

Laid to Rest: September 28, 2001

FALLEN HEROES

Gregory Sikorsky

Firefighter

Engine 46 (Working in Squad 41)

Memorial Service: October 20, 2001

Stephen Siller

Firefighter

Squad 1

Memorial Service: October 3, 2001

Stanley Smagala, Jr.

Firefighter

Engine 226

Memorial Service: October 13, 2001

Kevin Smith

Firefighter

Haz-Mat Co. 1

Memorial Service: October 6, 2001

SEPTEMBER 11, 2001

Leon Smith, Jr.

Firefighter

Ladder 118

Memorial Service: November 3, 2001

Robert Spear, Jr.

Firefighter

Engine 50 (Working in Engine 26)

Memorial Service: November 10, 2001

Joseph Spor

Firefighter

Ladder 38 (Detailed to Rescue 3)

Memorial Service: October 13, 2001

Timothy Stackpole

Captain

Division 11

Laid to Rest: September 24, 2001

FALLEN HEROES

Lawrence Stack

Battalion Chief

Battalion 50

(Detailed to Safety Battalion 1)

Memorial Service: December 8, 2001

Gregory Stajk

Firefighter

Ladder 13

Memorial Service: October 5, 2001

Jeffrey Stark

Firefighter

Engine 230

Memorial Service: October 1, 2001

Benjamin Suarez

Firefighter

Ladder 21

Memorial Service: November 2, 2001

SEPTEMBER 11, 2001

Daniel Suhr

Firefighter

Engine 216

Laid to Rest: September 17, 2001

Christopher Sullivan

Lieutenant

Ladder 111

Memorial Service: October 2, 2001

Brian Sweeney

Firefighter

Rescue 1 (formerly Squad 288)

Memorial Service: November 28, 2001

Sean Tallon

Firefighter

Ladder 10

Laid to Rest: November 2, 2001

FALLEN HEROES

Allan Tarasiewicz

Firefighter

Rescue 5

Laid to Rest: November 1, 2001

Paul Tegtmeier

Firefighter

Engine 4

Memorial Service: October 13, 2001

John Tierney

Firefighter

Ladder 9

Memorial Service: October 6, 2001

John Tipping II

Firefighter

Ladder 4

Memorial Service: October 15, 2001

September 11, 2001

Hector Tirado, Jr.

Firefighter

Engine 23

Memorial Service: November 8, 2001

Richard VanHine

Firefighter

Squad 41

Memorial Service: September 29, 2001

Peter Vega

Firefighter

Ladder 118

Memorial Service: October 6, 2001

Lawrence Veling

Firefighter

Engine 235

Memorial Service: December 1, 2001

FALLEN HEROES

John Vigiano II

Firefighter

Ladder 132

Missing

Sergio Villanueva

Firefighter

Ladder 132

Missing

Lawrence Virgilio

Firefighter

Squad 18

Laid to Rest: September 20, 2001

Robert Wallace

Lieutenant

Engine 205

Memorial Service: October 27, 2001

SEPTEMBER 11, 2001

Jeffrey Walz

Firefighter

Ladder 9

Memorial Service: October 6, 2001

Michael Warchola

Lieutenant

Ladder 5

Laid to Rest: September 21, 2001

Patrick Waters

Captain

Special Operations

Laid to Rest: October 5, 2001

Kenneth Watson

Firefighter

Engine 214

Laid to Rest: November 9, 2001

FALLEN HEROES

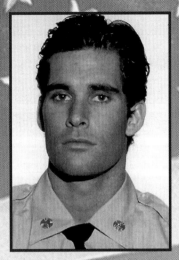

Michael Weinberg

Firefighter

Engine 1

Laid to Rest: September 17, 2001

David Weiss

Firefighter

Rescue 1

Memorial Service: September 30, 2001

Timothy Welty

Firefighter

Squad 288

Memorial Service: November 11, 2001

Eugene Whelan

Firefighter

Engine 230

Memorial Service: September 27, 2001

SEPTEMBER 11, 2001

Edward White

Firefighter

Engine 230

Missing

Mark Whitford

Firefighter

Engine 23

Memorial Service: October 4, 2001

Glenn Wilkinson

Lieutenant

Engine 238

Laid to Rest: September 17, 2001

John Williamson

Battalion Chief

Battalion 6

Laid to Rest: September 20, 2001

FALLEN HEROES

David Wooley

Captain

Ladder 4

Memorial Service: October 6, 2001

Raymond York

Firefighter

Engine 285

Laid to Rest: September 15, 2001

"I will die only when love dies, and you will not let love die."
Part of a sermon from one of the many memorial church services following the disaster.

SEPTEMBER 11, 2001

Photos by Time, Inc.

AFTERWORD

By John W. Norman III
Battalion Chief, Special Operations Command, FDNY

THE ATTACK ON THE CITY OF NEW YORK ON SEPTEMBER 11, 2001 WAS A SERIOUS WOUND TO THE FIRE DEPARTMENT. THREE HUNDRED FORTY-THREE MEMBERS OF THE FDNY WERE SUDDENLY AND VIOLENTLY TORN FROM US. THE NUMBER IS STAGGERING IN ITSELF, BUT THE REAL TOLL IS THE HUMAN SIDE OF THIS TRAGEDY.

FALLEN HEROES

"We are only passing through .... We are the guardians and custodians of a 100-year tradition." Deputy Commissioner William Feehan

Photos courtesy of FEMA

Photo by Time, Inc.

Each of the faces you have just seen, each of the brief vignettes you have read is just a fleeting glimpse at a life that is often among the most enviable on earth. From the youngest "probies" to the most senior chiefs, the lives of these men have many things in common, including a real zest for life that comes to those who face the realization that the "business" they are in comes with the risk that they may find that life taken from them. That is part of the life of a New York City firefighter. From 1865, when the FDNY was formed, until September 10, 2001, a total of 778 members made the supreme sacrifice. They are now joined by the 343 of September 11th. They will be joined by more, no doubt, in the future, all truly soldiers in the "war that never ends." I have had the privilege of serving with many of these men, creating friendships that require few words along the way, unspoken love. The kind of love that makes you willing to risk your life to try to save theirs. As Jesus Christ put it, "Greater love hath no man than that he lay down his life for a friend."

Now six weeks after the attack my thoughts of these men are not about the way they died, but about the way they lived. The word that comes to mind is "fully." They lived life fully. These

SEPTEMBER 11, 2001

Photos courtesy of FEMA

were men who lived and loved the life of a firefighter. They had their trials and tribulations, including how to make ends meet on the pitiful salary the city pays its firefighters the first five years on the job. They worked hard to support their families and they played hard, their concern for the welfare of their fellow citizens always present. They were not the type to walk away from a problem or a challenge, even one that didn't directly affect them. I think of how Firefighter David Weiss stopped his car on an elevated highway after seeing a car plunge off a road into the East River below. Dave turned to his girlfriend and said, "I'll be right back." He then jumped out of the car, climbed over a concrete wall, walked along a steel I-beam, shimmied down another steel column to reach the ground 30 feet below, dove into the swift current of the river, swam out to the car, dove down, and pulled the driver out of the vehicle. He swam to the seawall with the victim in tow, administering mouth-to-mouth resuscitation while awaiting help to pull the stricken man up to the roadway. Dave never once stopped in his rescue efforts to find out the victim's age, sex, race, religion, occupation, or political philosophies–he just went to help.

Photo by Time, Inc.

I think of one of the greatest acts of bravery I have ever witnessed. Rescue 2 was operating at a fire in a knitting mill on the first floor with what appeared to be apartments above. Mike Esposito was part of the "floor above" search team, charged with searching the floors directly over the fire for victims. Larry Weston and Dennis Mojica, the chauffeur, were assigned to assist the first due Engine Company with getting its hoseline to the seat of the fire. A very heavy fire was resisting their advance and making conditions on the floor above unbearable. Dennis and the captain of the engine company knew that a serious situation existed and conferred on how to proceed. In the best tradition of an aggressive engine, they did what all good enginemen do, put their heads down and backs into the line and pushed forward deeper and deeper into the inferno. On the second floor, we were taking terrible punishment, our hoseline was hitting fire everywhere, but no matter where you turned fire kept coming up from below. So I told Mike Esposito to hold fast and I would go downstairs to get them to put the fire out under us. I had just started to crawl in on the first floor, following Dennis' line, when a major collapse occurred right in front of me. Something very heavy had

September 11, 2001

"You can't sit back and say, 'Poor me. Poor me.' You have to learn from this incident and move forward."

Jim Ellison, FDNY

just fallen through the second floor and the whole area in front of me filled with a sheet of flame. Immediately I started calling on the radio for Dennis or Larry. I made several calls, but got no answer. I started to crawl back toward the front of the store to get help, and a hoseline to knock down the fire. As I got to the doorway, Mike Esposito bumped headfirst into me. He asked what happened and I told him that there'd just been a major collapse and I thought Dennis, Larry, and the others may have been trapped. I told him to go get a hoseline and follow me in, but Mikey said, "No, Lieu, you go get the line. The chiefs won't listen to me, but if an officer comes out, they'll listen to you." Then he said, "I'm going to get those guys," and disappeared into a wall of flame like something out of a Hollywood special effects scene. As I said, the bravest thing I've ever seen. The best part is that he did get to them, and they were all okay on the far side of the collapse. They had located a back door and made their way out into a rear yard. None of their radios were working by the time they got to safety and were unable to answer my calls, but they were safe. Mikey literally climbed into hell, getting a number of first and second degree burns in the process, because there were men trapped. That's the kind of men that went into those twin towers on September 11th…heroes.

FALLEN HEROES

Photo by The Associated Press

It is not possible for me to write a story like those two about every one of the 343 members of the Fire Department that perished that terrible Tuesday. There is not enough room in this book for all of those stories, for one thing. I could tell you many of the great stories about Pete Ganci, Ray Downey, or Terry Hatton. They were all awe-inspiring men, each in their own way. I could also write about Joe Angelini, of Rescue 1, the senior firefighter in the Department, with over 40 years on the job, or of his son, Joseph Angelini, Jr., of Ladder 4. Both men were lost that day, leaving a gaping hole in the fabric of our Department, big enough to drive a tower ladder through. Yet I can't tell you about all 343 of these brave men because I didn't know them all. The FDNY is a huge job, more than 11,000 fire officers and firefighters, and 2,500 emergency medical technicians and paramedics. It is simply not possible to know every one of them personally. I didn't know many of the firefighters from the 57th Battalion in the heart of

"There is a vast difference in viewing the disaster scene in newspaper photos and on television screens. There were eight- to ten-story-high piles of rubble that appeared to be never-ending. They stretched for at least five blocks in every direction." Alan DeRosa, Hazardous Materials Officer, East Rutherford (NJ) Fire Department

September 11, 2001

"You must not lose faith in humanity. Humanity is an ocean; if a few drops of the ocean are dirty, the ocean does not become dirty." Mahatma Gandhi

Photo courtesy of FEMA

Photo by The Associated Press

Brooklyn, or the 46th Battalion in north Queens, men who never imagined that they would be fighting for their lives in a Manhattan high-rise. But I know what they were like. They were the same as Ray, Pete, Terry, Mike, Dave, and Joey. They were like many of you and me. Guys who took their sworn oath to "protect life and property" seriously. They, like thousands of Americans before them, died doing their job, because it was their job. No one else was going to go in there and pull people out of that flaming hell. It was going to take firefighters, and they were the ones to do it. Not one stopped at the front door and said, "I'm not going in there, it's too dangerous." The legions of civilians that they helped save, tens of thousands of people, tell over and over how the firefighters guided them down the stairs in their escape, using their flashlights in the darkened stairways, carrying those too seriously injured or infirm to proceed on their own, offering direction and words of encouragement to the exiting throngs, all the while plodding upward with their burden of protective clothing, masks, and tools. Stopping only to tend to a seriously injured person or momentarily for a brief respite. Upward to the 30th, 40th, and 50th floors, the youngest probies helping shoulder some of the burden of the older, more seasoned veterans, they continually surged upwards. Staff chiefs like Donald Burns and Gerry Barbara organized teams of companies, assigning groups of these teams to individual Battalion Chiefs with assignments to search groups of floors, "You take 30 to 40! You take 40 to 50, Joe, try to get as high as you can!" Upward they went…heroes.

While I didn't know all these brave men personally, there is a saying that is repeated in firehouses throughout the city that epitomizes their lives. It states "A firefighter commits his greatest act of bravery when he takes the oath of office. Everything else is part of the job." In the weeks after the attack, the spirit of the job was battered, but not broken. Firefighters who had worked for hours digging through the smoldering pile or searching through the labyrinth-like below grade area that extends six stories below ground, wrote messages of encouragement to the others that would follow them, using their fingers to scrawl through the layers of dust and grime that cling even to vertical glass surfaces. Among the most heartwarming – "FDNY: still the greatest job on earth!" By the end of the second week printed

Photos by Time, Inc.

"Only a life lived for others is worth living." Albert Einstein

September 11, 2001

"It is not the critic who counts, not the man who points out how the strong man stumbled, or where the doer of deeds could have done them better. The credit belongs to the man who is actually in the arena; whose face is marred by dust and sweat and blood; who strives valiantly; who errs and comes short again and again; who knows the great enthusiasms, the great devotions, and spends himself in a worthy cause; who, at the best, knows in the end the triumph of high achievement; and who, at worst, if he fails, at least fails while daring greatly, so that his place shall never be with those cold and timid souls who know neither victory nor defeat." **Theodore Roosevelt**

posters bearing the message had sprouted everywhere on the site. The job is still great. And the rebuilding process is already under way. Just weeks after the tragedy, 305 new probationary firefighters were sworn in, joining their "brothers in battle" and 168 fire officers were promoted a week after the attack, to try to fill the void left by the tragedy, each of them motivated to do whatever needed by the memory of those who preceeded them. In my case, I follow in the footsteps of one of the job's greatest officers, Chief of Special Operations Ray Downey. Ray had been my mentor for over a dozen years, giving me an opportunity to serve as his Lieutenant in Rescue 2 when I was a young officer surrounded by some of the "giants" of the job (legends like Pete Bondy, Lee Ielpi, Richie Evers, and Larry Gray). Ray believed in me, and his word was good enough for those guys. Five years later Ray was the Chief of Rescue Services and there was an opening for the Captain of Rescue Co. 1. With all the talent he had available to choose from, guys like Patty Brown and Jim Ellson, Ray and Chief of Operations Donald Burns chose me for the assignment. I will always be thankful for his confidence and inspiration, and strive to rebuild his shattered Special Operations Command in his honor. That's the way all of these men would have it. Don't look back, keep going forward–onward and upward!

"**Prayer has comforted us in sorrow, and will help strengthen us for the journey ahead.**"
President George W. Bush

Photos by Time, Inc.

SEPTEMBER 11, 2001

THE FIREMAN'S LAST CALL

Photo by The Associated Press

By Jennifer Narcisco

September Eleventh
The city was lit bright
The Twin Towers standing side by side
The Statue of Liberty standing with Pride
A plane struck the north tower
As people began to scatter
Fire lit the city's sky
As people began to cry
In between an hour
Another plane hit the South Tower
Now both towers were burning with fire
As people still begin to scatter
The North Tower suddenly becomes weak
And falls into pieces all over the streets
The South Tower wasn't any stronger
It couldn't hold any longer
The building gave in
As it collapsed, caving in
Black smoke filling the streets and sky
Had millions wondering why
Firefighters putting out the fire
As rescue workers try to find survivors
Thousands are lost
Thousands still missing
Some not badly injured
Some in critical condition
It seems like a nightmare that will never end
Thousands of people are now dead
It started off with a brightly lit sky
Which now turned into a world with fire in their eyes
And a world that lost thousands of lives

Jennifer Narcisco
Age 14
Daughter of Vincent Narcisco
1st Asst Chief, Valley Cottage Fire Dept, NY

FALLEN HEROES

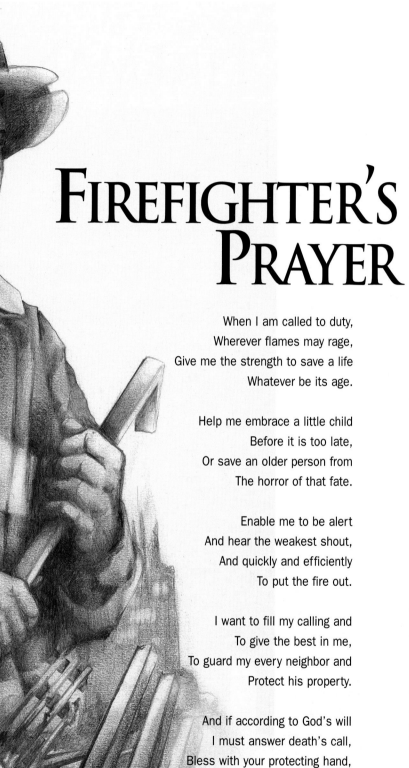

FIREFIGHTER'S PRAYER

When I am called to duty,
Wherever flames may rage,
Give me the strength to save a life
Whatever be its age.

Help me embrace a little child
Before it is too late,
Or save an older person from
The horror of that fate.

Enable me to be alert
And hear the weakest shout,
And quickly and efficiently
To put the fire out.

I want to fill my calling and
To give the best in me,
To guard my every neighbor and
Protect his property.

And if according to God's will
I must answer death's call,
Bless with your protecting hand,
My family one and all.